ROUGH
SEAS

ROUGH SEAS

SEAS THE LIFE OF A DEEP-SEA TRAWLERMAN

JAMES GREENE

The
History
Press

First published 2012

The History Press
The Mill, Brimscombe Port
Stroud, Gloucestershire, GL5 2QG
www.thehistorypress.co.uk

Reprinted 2017

British Library Cataloguing in Publication Data.
A catalogue record for this book is available from the British Library.

ISBN 978 0 7524 6453 4

Typesetting and origination by The History Press
Printed in Great Britain by TJ International Ltd, Padstow, Cornwall

CONTENTS

ACKNOWLEDGEMENTS

Amy Rigg, Emily Locke, Kerry Green and The History Press for all their help; Austin Mitchell for all his help; the late Dolly Hardy who did a lot of hard work to get fishermen compensation; Eric Fearnely, Photographers, Grimsby; Grimsby Central Library; Grimsby Heritage Centre; Innes Photographers, Hessle Road, Hull; Jamie Macaskill, *Hull Daily Mail*; Janeen Willis & Fred Goodman, photographic collection; Jim Porter, administrator at Fleetwood-Trawlers.info; Lyne Asgha, Fleetwood Museum; Mary Houghton for 'The Last Voyage' by her father in chapter 12; excellent photographs from Memory Lane, 528 Hessle Road, Hull; Michelle Lalor at the *Grimsby Telegraph*; my daughter Alison for all her help on the computer; Steve Farrow for the use of his wonderful paintings; Susan Capes at the Hull Museum & Art Gallery; the US Coastguard; William H. Ewen, Steamship Historical Society of America.

GLOSSARY

Aluminium floats	For keeping the top part of the net off the bottom of the seabed.
Becket	A short piece of rope with an eye spliced on each end.
Bobbins	Iron bobbins 21-24in in circumference for keeping the lower part of the net on the seabed.
Brassy/Decky	A deckhand learner.
Busters	Bread buns.
Fore Peak	A compartment in the bows of the ship for stowing the anchor chain.
Gilson	A steel wire that goes from the port side deck through a running block at the top of the mast and down to the starboard side on which a hook is shackled and used for heaving heavy gear about the deck.
Leeside	The leeside is the sheltered side. If you sit behind a wall sheltering from the wind you are then under lee.
Lumpers	The men who work ashore to land the fish.
One-ended job	The trawl is towed along the seabed on two wires. If one wire parts all the weight of the gear is on one wire – the one-ended job – and if that wire parts all the gear is lost.
Paralysed	When the net is torn to shreds.
Port side	Left-hand side of a ship.
Runner	The man who signs the crew on the ship; he is also responsible for her getting to sea on time.
Starboard	Right-hand side of a ship.
Tackle	A purchase wire similar to the Gilson for heaving really heavy weights about.
Trawl	The net that holds the fish.
Yoyo derrick	A derrick attached to the mast to heave anything out board.
MT	Motor trawler.
Shore gang	The men employed to work on getting a ship ready for her next trip.
ST	Steam trawler.

FOREWORD

I am delighted to welcome Jim Greene's excellent autobiography because it's not only his own life story, but a marvellous insight into the industry in which he served. We won't see the like of either again so it's good that Jim has not only told the tale of fishermen like him who spent most of their lives in peril on the sea, but gives us a fascinating picture of the rise and fall of our once great fishing industry. Fishing shaped the lives and built the fortunes of the three great English fishing ports of Grimsby, Fleetwood and Hull, but was almost totally destroyed by the world trend to 200-mile limits and by our own inability to rebuild fishing in our own rich waters thanks to a Common Fisheries Policy which gave equal access to a common resource, meaning that our rich fishing grounds were open to the French, the Dutch, the Danes, and others all managed not by us for our purposes but by the European Commission doling out paper fish in political decisions taken in Brussels.

Jim's life brings home the drama, the dangers, the highs and the lows of the lives of the thousands of fishermen who toiled in our distant water fishing industry up to the time when it suddenly found itself with nowhere to fish, and that story deserves to be told and remembered before the last generation of distant water fishermen fades into history, as most of their industry has already done. In Jim's day Grimsby boasted of being the World's Premier Fishing Port, a title disputed by Hull but rightfully ours because, though Hull caught more fish and regularly won the annual Silver Cod award, Grimsby caught quality rather than quantity and had a bigger and more diversified fleet fishing middle and near waters as well as Iceland, Greenland, the Faroes, Norway and even Canada.

So Jim is describing not just an industry, which was very profitable for the men and even more so for the big powerful owners who ran it, but a way of life unique to the three ports where fishermen lived in tight fishing communities of terraced housing clustered round the docks, sons followed fathers, as Jim did his, and wives raised the families while absentee fathers fished for three weeks away, then home for perhaps three days.

The men faced the restless waves and the distant waters but spent much of their time when ashore in the rough, tough world 'Down Dock' from which men, money and life spilled out into the surrounding streets, particularly in Freeman Street, Grimsby, with its pubs, shops, markets, training facilities and schools. Down Dock was the focus of fishing life and an almost tribal existence. The offices, the engineering, provisioning, landing marketing and auctioning of the fish all carried on there and the merchants' and owners' premises clustered round the market from which the precious fish were despatched by rail, up to Dr Beeching's railway cuts, then by road out to the waiting world.

Down Dock was a tough, male, insular world somewhat separated from the rest of Grimsby, as was Hessle Road in Hull and the fishing streets of Fleetwood. Fishing made the towns rich but in return the respectable burghers looked down, almost askance, at the industry and the men who'd generated the wealth of their towns. Down Dock was another world, one where fishermen worked, lived and strutted the nearby streets like kings, even though, in fact, they were lashed to their industry like seaborne serfs by the powerful cartel of fishing vessel owners whose word was law.

It's good and right that that way of life should be written up for history and Jim Greene has done a marvellous job in recreating the industry by telling his own life story simply and honestly and without pretension or stylistic tricks for the future generations who'll never see its like again. With all its toughness, tragedies and terrors in distant waters our big, English fishing industry has gone, leaving perhaps a score of middle water vessels in Grimsby, a few freezers in Hull and a small remnant, some Spanish owned, in Fleetwood. The fishing communities have been broken up, their continuities and lifestyle destroyed. Believe it or not, it's now difficult to recruit fishermen in the old ports because so few youngsters want to go into an industry to which most once flocked. Jim never wanted his own son to go into the industry to which he'd devoted his own life.

The death of fishing was sad, even shocking, because a once powerful industry was scuttled. Its vessels were sold off, sent to Spain, Africa or anywhere which could pay, or scrapped. Its owners, well compensated by the government, put the money in their pockets and scattered, leaving the towns that had made them and taking their money with them. The men were thrown onto the scrapheap with neither compensation nor redundancy because these slaves of the sea were viewed as 'casual'. Neither was paid until a tribunal case in 1994 proved that fishermen were in fact employed, not the casual workers this almost captive workforce had (unbelievably) been seen as in 1976. Then only thirty years after the death of the industry and after a long struggle by the British Fishermen's Association, by Dolly Hardie and by the port MPs, the Labour Government paid compensation for the loss of jobs and livelihood, a compensation which is only now being grudgingly and unfairly finalised. No other group of workers had such a hard, dangerous life. None has been as badly treated.

There is no point in me telling that part of the story over again. Jim Greene tells the whole story brilliantly in his own words through the narrative of his life, starting in Fleetwood, moving upmarket to Grimsby, fishing first as deckie learner, then mate and finally skipper. He was one of the youngest, too, but thank heavens his tough life has been remembered and preserved forever as it deserves to be. Enjoy.

Austin Mitchell
MP for Grimsby

CHAPTER I

HAYBURN WYKE, 1938

It was a cold, stormy night in March 1938, the month and year I was born. Somewhere off the west coast of Scotland the Fleetwood coal-fired steam trawler *Hayburn Wyke* was dodging, her bow ploughing into the teeth of a storm. Fishing operations had been suspended due to the bad weather. All crew were sleeping except for two men on watch in the wheelhouse and the chief engineer and his firemen working hard to shovel coal into the furnaces to keep fires going and the engine running smoothly.

In the wheelhouse the watch kept a sharp lookout for other ships while the bosun studied the weather. An ice cold wind whistled through the wheelhouse, in darkness except for a small light that lit the compass. The only other light came from the depth sounder flashing at regular intervals. It was noisy; the windows rattled in the gale and the ship was constantly battered by heavy spray. The wind ripped through the rigging, howling like some tortured creature lost in the night.

The man at the wheel struggled, constantly fighting as the trawler kicked and pulled, desperate to keep the ship's head to wind. The night was clear, visibility good with a bright moon shining. Briefly the islands of St Kilda stood proudly on the horizon bathed in the light of the moon, only to disappear again as the next wave loomed and blotted them out. Every now and again a large sea would build up in a threatening manner, to drop away to nothing as it reached the bow.

The bosun watched the following sea; he knew it posed a serious threat. The helmsman had already spotted it and clung to the wheel. The sea was now a sheer wall of water starting to break; they watched in horror as it curled over the bow, sending tons of water crashing onto the deck. An almighty bang was heard from somewhere aft.

The wheel was wrenched from the helmsman's grip, spinning crazily like some out of control fairground ride. He made another grab for it, trying to regain control, but was too late: it was slack in his hands. The ship was rapidly turning to starboard, broadside to the heavy seas.

Now helplessly out of control with a smashed rudder, the bosun was first to react. He ran to the telegraph to stop the engine. On hearing the commotion the skipper, sensing something was seriously amiss, leapt from his bunk. Grabbing his Guernsey, muffler and cap he hurried to the wheelhouse. He contacted nearby trawler *Dinamar* requesting she stand by until the weather eased down or they could begin repairs. The other ship agreed to do all in their power to assist.

Several hours passed, the weather improved slightly and the crew worked relentlessly to rig up temporary steering gear enabling them to get closer to St Kilda and more sheltered waters. As they headed for calmer water the skipper and mate discussed a more permanent repair idea: to hang one of the trawl doors under the counter of the stern using chains and shackles. A warp from each side of the trawl winch would run down both sides of the ship and be shackled to the trawl door. The steam winch could then be used to control the emergency rudder.

The first job was to pump tons of water down the forepeak to sink the bow and lift the stern out of the water, enabling the crew to work underneath. A difficult task indeed; the ship pitched and rolled in the heavy seas and as waves broke over them she drifted further from land and shelter. The lifeboat was launched with the skipper, bosun and two deckhands on board. There was the danger that the lifeboat would get caught under the stern, capsize and throw the occupants into the sea.

One brave deckhand offered to swim round the stern to secure a shackle to the smashed rudder. The skipper wouldn't allow it, it was just too dangerous.

With great difficulty they managed to heave the trawl door over the side and suspend it on a chain over the stern. The door, made of wood and steel, weighed around 1.5 tons. Several times it swung too close to the lifeboat causing great concern to all.

Once in position it was heaved tight into the ship's stern and secured with chains and shackles. The warps from the winch were shackled to the door. By heaving and lowering the warp they managed to steer the ship.

Sunday lunchtime and work was eventually finished. The lifeboat was back on board and lashed down, the decks cleared away and the trawl doors stowed. However, during operations the derrick had swung across the deck and smashed into the skipper, injuring his shoulder.

The crew had worked non-stop for thirty-six hours at the mercy of the gale to get the job done. Despite being exhausted, wet, cold and with aching limbs both skipper and crew felt a sense of pride at what they had achieved.

By standing at the winch on the windswept deck, heaving and slackening the warps as required, they were just able to steer. Twice on the way back to

Fleetwood the warps parted. The ship drifted helplessly towards land as warps were spliced (repaired). For 300 miles they struggled to get back home.

A tug stood by to tow them in when they reached Wyre Light just off Fleetwood. Even then worries weren't over; the tow rope parted three times. Each time they drifted up Morecombe Bay before the tug could retrieve them. With the exception of the skipper who was twenty-eight, not one crew member was more than twenty-four years old.

It was seen as a magnificent feat of seamanship. Awards were made to skipper and crew. It was thought this was the first time a trawler had been able to effect repairs of this nature in such a storm, without having to be towed home by another vessel.

The skipper received the grand sum of £36 10s 6d from the insurance company.

That skipper was my dad.

CHAPTER 2

WAR YEARS, 1939 TO 1945

The Second World War broke out the following year, September 1939. Dad joined the Royal Naval Reserves (RNR) as acting skipper lieutenant. He was injured in action three times during his first year at war. In the English Channel on 25 November 1940 he was on the HMT *Kennymore* when she was blown up and sunk. He was rescued with the other survivors by HMT *Conquistador*. On her way, the *Conquistador* was in collision with HMT *Capricornus*, which suffered damage to her bow, but *Conquistador* was severely damaged, sank and again Dad was rescued, I assume by the *Capricornus*. He was taken into port where he spent six months in hospital at Sheerness.

On 7 December 1940 the *Capricornus* was herself blown up by a mine and sank.

As Dad had spent several years as skipper of the *Hayburn Wyke* he put in a request to return as skipper on her but the Royal Navy turned him down. The *Hayburn Wyke* was torpedoed and sunk by a U-boat while anchored off Ostend on 2 January 1945. In the Civil Honours List Dad was awarded the Distinguished Service Cross (DSC) for bravery in the face of enemy action.

As children we lived in a three-bedroomed house, middle of a row of three, on Broadway, the main road between Fleetwood and Blackpool, with the added luxury of a bathroom and toilet upstairs.

At the outbreak of war we were issued gas masks. They came in a cardboard box with a shoulder strap and we were instructed to take them everywhere with us. They even manufactured them for small babies. When I first tried mine it wasn't a pleasant experience, let me tell you! It was very claustrophobic; I started breathing fast and felt like I was suffocating. I panicked and pulled it off. Once I got used to it it wasn't too bad, but I never put it on unless told to!

The war years were very hard for my mother. At the outbreak of the war there were four of us. Betty was the eldest at six years old, Odette was four,

Billy two and I was just a year old. In 1940 Pat was born followed in 1942 by Peter. It must have been difficult for my mother with six children to feed and clothe and everything scarce or rationed. It got so bad that one time Betty was sent to Grandma's in Preston as food was so short.

At one stage Odette was found to have diphtheria and was sent to hospital in Moss Side near Manchester. The diphtheria virus being so contagious, all the rooms in the house were treated – systematically sprayed then sealed up, how long for I really can't recall.

Another time Pat contracted bronchial pneumonia. She was in hospital for almost a year, very seriously ill. I believe she was the first person in Fleetwood to receive penicillin.

The lowest point for us all was when our mother went into hospital. She suffered with leg ulcers and they'd become so bad she had to be admitted. The authorities I think considered taking us into care but Grandma saved the day. She came to Fleetwood to look after us so we could all stay together at home.

During the war three bombs were dropped on Fleetwood. One aimed at the ICI plant but missed. I'm surprised there weren't more because the ICI plant was one of the biggest in the country.

We always looked forward to Dad coming home; he always brought us a present and gave us a florin and an orange each, a rarity back then. He'd bring home Turkish delight too but I wasn't keen on it! He spent time in Egypt during the war and he'd bring home lovely jewellery for my mother.

We thought we had it bad but we got off lightly compared to the big towns and cities like London, Coventry and the like – even Hull and Grimsby got it bad.

The Grimsby dock tower started its working life in 1852, officially opened in 1854 by Queen Victoria. In 1931 a large earthquake on the Dogger Bank measuring 6.1 on the Richter Scale shook the tower but it remained structurally sound. During the war it survived the bombing of Grimsby as it was a useful landmark to the Luftwaffe, being a reference point to fly due west to Liverpool. On their way back to Germany any bombs left were dropped on Grimsby and Hull before crossing the North Sea back to Germany. If an aircraft was damaged there was no way they wanted to crash land on their own soil with bombs on board.

So we had it easy compared to others. My then future wife lived on the Thames in Erith and some of the stories she can tell are terrifying. They were bombed repeatedly night after night as they lived close to Woolwich Arsenal, a huge target. For years after she was absolutely terrified of thunderstorms.

We were lucky as my dad returned home safe from the war. I knew a lot of boys and girls at school whose dads didn't, and they were the ones that had it hard after the war. I hardly remember much more about it all as I was too young.

CHAPTER 3

COMITATUS, 1946 TO 1952

After the war Dad returned to Fleetwood and took command of the Fleetwood coal-fired steam trawler *Comitatus*. Registered GN 39 and built in Beverly in 1919, she was 120ft in length and weighed 108 tons.

I was eight years old; I'd never given much thought to going to sea, although my friends and I were mad about trawlers and fishing. It came out of the blue when Dad said he was going take Bill and me to sea with him. It was the best present I could have wished for. It was like a birthday and Christmas rolled into one. Talk about being excited; I was out of that house like a shot to tell my mates. The down side was it wouldn't be on his next trip as we'd not broken up from school for the summer holidays. We had to wait a whole two weeks!

Mother took us into town to get warm clothes and wellingtons. Dad was still at sea and I kept pestering her to find out when he was coming in. She said I was getting on her nerves, I couldn't think why!

I'd go down to the prom at Fleetwood when Dad came in, or the Jubilee Quay where the Isle of Man boat was berthed. You could shout and wave to the crew from there as they came so close to the quay. Dad would give us a blow on the ship's whistle. You could tell when it was coming as a plume of white steam came from the whistle on the funnel seconds before you heard it, a real deep sound you could feel vibrating in the pit of your stomach.

The night before we sailed it was early to bed, a complete waste of time as there was not much sleep to be had. It seemed like the longest night of my life. The taxi was ordered for 5.00a.m. that morning. We were up and ready – the taxi was late. I started to think they'd forgotten us. The taxi arrived at 5.15a.m. Finally we were off! I could hardly contain my excitement.

We stopped off on the way to pick up Bill's friend. We arrived on the docks at 5.45a.m. There was a lot hustle and bustle going on, taxis arriving, doors slamming as crews arrived for sailing. There were about seven or eight ships due to sail that morning. My dad's friend George Beech was skipper on one of them, the *Buldly*. Also sailing that morning was the *Wyre Mariner* and the *Bridesmaid* on which my granddad was bosun.

The ship's runner had his work cut out. His job was to make sure all the crews were down and the ship's sailed on time. When the crews were down he would inform the lock gates. The first ship ready was the first ship to sail. Our ship was ready when we arrived at the dockside; the shore gang had been down since midnight to get the fires going ensuring there was plenty of steam in the boilers ready for sailing.

The sun was just rising; there was a light easterly breeze blowing with a distinct chill that sent shivers through my body. The air was filled with the smell of steam and coal smoke as it drifted lazily about the docks on the light breeze. At around 6.00a.m. the dock gates opened and we climbed aboard. The ship's ladder was pulled on to the quayside. We were the fourth ship to sail that morning.

We had to wait until the ship's name was called out on the loud hailer at the lock gates. After what seemed an eternity, *Comitatus* was called. The crew wereon the whale back, Dad and mate on the bridge. Two men worked the after ropes.

'Let go forward,' Dad shouted.

The ship's runner threw the ropes off the quayside bollard and let them slide into the water. The men on the bow pulled them on board and they were stowed away ready for sea.

The ship's telegraph jangled with a hiss of steam, and you could hear the thump of the steam engines as they sprang into life. The whole ship trembled as we came slowly astern; the after mooring rope came tight as it took the weight of the ship. The ship's bow started to move away from the quayside. When the bow had cleared the two ships moored in front, Dad shouted, 'Let go aft!'

The after rope was slipped then pulled back on board. We slowly made our way across the dock to the lock gates.

It was a smooth operation as the trawler edged its way through the lock gates and out into the River Wyre, edging past the notorious Tigers Tail, a sandbank on our starboard side. Many ships ran aground here over the years, especially in fog.

The ship's telegraph jangled again, and this time the engine really came to life as the chief engineer notched up to full speed. The whole ship shook from stem to stern; huge plumes of black smoke billowed out of the funnel and drifted across the channel.

The engine settled down to a steady rhythm as she picked up speed. There was a great excitement about it all; it was just brilliant. From that moment I was well and truly hooked. This was the start of my life at sea and the beginning of a great journey.

I sailed with my dad between 1946 and 1952. During this time I did four-teen trips on the *Comitatus* and one trip on the *James Lay*. Each trip was around fourteen to fifteen days. I'll tell you about the things I remember hap-pening during these pleasure trips.

We were fortunate that first time. The weather was hot and the sea was calm, just like a millpond. I thought maybe it would be like this all the time, but boy I soon knew different. By the time I left school at fifteen I'd experi-enced gales and storms more fearsome than I could ever have imagined.

That first trip was like a big adventure. It was all new to us. The weather was calm, the sun shone. We left the dock and entered the River Wyre. We sailed passed the Jubilee Quay were all the small inshore boats were tied up. My mother stood on the seafront to see us off. We waved and shouted until we couldn't hear what she was saying anymore. We sat on the casing and watched as we sailed out of the river, passing the Isle of Man boat moored up to the quay. We passed the Pharaoh's Lighthouse, Euston Hotel and Fleetwood Pier. We then made for the Wyre Light.

Dad called us to the bridge to inform Bill and me we were to sleep in his berth. Bill's friend had a bunk aft with his uncle. The first day was spent exploring the ship and getting to know the crew.

Dad had a few rules about what we could and could not do. The main rule was not to go on deck at all while the nets were being hauled up from the seabed or as they were being lowered away.

We fished on the hundred fathom edge to the north-west of Scotland, all the way to the north-east along the hundred fathom line towards Muckle Flugga, the most northerly point of the British Isles. We were mostly after hake, the prime fish for Fleetwood at that time.

The journey from Fleetwood to Cape Wrath on the north coast of Scotland is one of the finest you can do. The scenery down the west coast and through the Sounds of Mull is breathtaking, especially in late summer when the heather is in full bloom, but even in the depths of winter when the snow covers the hills and glens.

The crew were all good men. The mate, a chap called Nelson Rogers, sailed with my dad for a long time but my favourite by far was the aptly-named Albert Cook, the ship's cook. He was great, he used to look after us, nothing was too much trouble for him and he was an excellent cook. Every morning Dad sent us aft with a bucket. Albert would fill it with hot water and we'd stand on the after deck to have a strip down wash. When we were done Albert would have a steaming mug of hot tea ready for us.

'Now lads what do you want for breakfast? How about a couple of boiled eggs?' he'd ask.

'Good idea Albert,' we'd reply.

Around ten o'clock in the morning the stove was littered with pots and pans all boiling away full of spuds, cabbage and other stuff for dinner. This never bothered Albert; he'd just toss three or four eggs into the pan of soup. While the eggs were boiling he would butter a couple of busters (fresh baked bread buns). They were the best. They made for great egg sandwiches!

I'd help Albert in the galley where I could, washing up or peeling spuds. During bad weather I had to stay out of the galley in case anything hot got thrown off the stove. In the afternoons when Albert was having his nap I made sure his coal locker was filled up and the fire didn't go out.

At the end of the trip Albert baked a dozen busters for me to take home. I also took home pepper. Just after the war things were rationed and pepper was scarce so I'd do my bit to help Mother. Often on the trip home the ships called into Southern Ireland. Lots of things in Southern Ireland were not rationed like at home. It was not unknown for a crew member in need of a new suit to go away in an old one and come home in a brand new one.

Albert was on the *Comitatus* all the time I sailed with Dad. I can't recall any other cook being there.

Another crew member I remember well was the deckhand learner Bill Barton, from Wharton just outside Preston. Aged about fifteen, he'd been with my dad for about a year since he left school. Deckhand learners were known as 'brassys'. This came about because around the turn of the century they'd wear a uniform ashore with brass buttons, similar to that worn by the Merchant Navy. In Grimsby he'd have been known as the decky learner and in Hull as the snacker. Bill and me got on well with him; we'd sit and talk for hours about fishing and all sorts. He was very keen, a good worker always willing to learn.

One trip we went into Londonderry. I believe it was a big naval base at the time. As we steamed up the Lough there where row upon row of German submarines, all captured or surrendered after the Second World War. We sat on the casing watching them go by. As we passed a couple of Royal Navy Frigates Bill jumped up and grabbed my arm, he was very excited.

'Look! Look!' he shouted, pointing to one of the ships.

'My pal's on her, I'm going to ask Dad if I can go and see him.'

Dad refused, he wanted Bill to take me to the pictures when we went ashore. Bill was very disappointed. After tea I washed in the bucket before Bill and I went ashore.

'Do you mind if we go round to the Naval base and see if I can see my pal?' Bill asked.

I was all for it, thinking that I would get on one of the navy ships. On the way we stopped to ask a policeman for directions. I couldn't believe my eyes when I saw he had a gun. When we arrived at the base the guard at the gate also had a gun. I had never seen a real gun before, let alone two in one day!

Bill enquired about his pal at the dockyard gate but we weren't allowed on the base for security reasons. I was a bit disappointed; I wasn't going to see the ships after all. The guard asked Bill the name of the ship his friend was on and he contacted the ship by telephone.

'You're lucky mate; he's getting ready to come ashore so if you hang about he'll be out shortly.'

Twenty minutes later Bill's pal showed up. I was introduced to him and Bill explained why he had me tagging along.

'Well I have to be back on board by 22.00 hours so we will go for a couple of pints and catch up,' Bill's pal informed us.

My dad thought we were going to the cinema but we were actually going to the pub and I had to promise not to tell. We set off down the street looking for a decent place to go; they all looked a bit rough. Bill decided they wouldn't be suitable for me. The third one looked quite respectable and after a bit of a discussion it was decided this one would do.

Bill's pal went in first. Bill started to follow but as he got halfway through the door somebody in the pub started to laugh. Bloody hell it was Dad! Panic stricken I grabbed Bill's arm.

'Bill,' I said in a whisper. 'Dad's in there don't go in.'

By this time his mate had disappeared into the pub. We had to hang around outside until he came out to see where we had got to and we made a quick retreat down the road. That was a bit too close for comfort. Had we gone in my dad would have gone mad, I wouldn't have been in Bill's shoes for the world!

The chief engineer (Webby) was with my dad all the time I did pleasure trips. I remember him well and he'd often come to our house when they were in dock. Webby must have been with my dad for about seven years. The second engineer was a chap they called Happy Harry. One trip everybody was running out of tobacco and cigarettes were in short supply. Some of the crew had run out altogether. Happy Harry still had tobacco left, but being the sort of person he was he'd see you gasping rather than give you a cigarette. Harry always saved his tab ends in a tin alongside his bunk, and it was nearly full.

A couple of the deckhands kidded me on to go and pinch them. At first I was a bit reluctant but the deckhands said he had loads of tobacco left. After a bit of persuasion I agreed. Once he was asleep I sneaked into his cabin, took all his tab ends and gave them to the deckies. They rolled them up into fags and shared them out. I must say I felt pretty good about it. After all, Harry had plenty of tobacco left, didn't he?

The crew was in the pound gutting fish. I was trying to gut a cod that was nearly as big as me but the cod was winning when Albert called out dinner was ready. For some reason my dad did not go for his dinner at 12.00 as he usually did but stayed on the bridge. I was sat in the galley with Albert. I was not allowed to have my dinner till 12.30.

As the crew was having dinner I could hear raised voices in the cabin, some sort of argument going on. 'Take no notice Jim, they are always falling out about something or other,' Albert said. At 12.30 the crew was back on the deck. Albert told me to go and get my dinner. I went down to the cabin. As I got to the bottom of the ladder Happy Harry started pointing and shouting at me.

'You!' he snarled. 'You pinched my fags for them bastards on the deck. I saw you sneaking out of my cabin when you thought I was asleep. I'm surprised at you Jim, I didn't think you'd do a thing like that.'

I could feel myself colouring up, I felt terrible and scared, nobody had ever shouted at me like that before. I was close to tears. With that my dad came down the ladder into the cabin. Bloody hell, now I was in real trouble!

'What's going on? What's all the shouting about?' said Dad.

Harry told him what I had done; I was expecting the worst, at least a clip around the ear hole. Dad looked at me, winked, then turned and said to Harry, 'That's bugger all to what his old man's done in the past. Don't worry about it Harry, come on the bridge later and I'll give you some tobacco.' My dad never mentioned it to me again, I don't know if he told the crew off. If he did they never said.

I attended three schools in Fleetwood, Flake Fleet Infants School, Chaucer Road Junior School and Bailey Secondary Modern School (nicknamed Bailey Borstal). While I was at Chaucer Road School the Fleetwood trawler *Goth* went missing. There were rumours going around that something was wrong, and a couple lads were off that day. It wasn't until after school we read in the *Fleetwood Gazette* that the she'd gone missing.

At 147ft, she was built in 1925 by Cook, Welton & Gemmel of Hull and Beverly for the Hellyer Bros Ltd of Hull. At the outbreak of the Second World War she was taken over by the Royal Navy for minesweeping. After the war she was sold to the Ocean Steam Fishing Company Ltd of Hull. Not until 1946 was she sold to the Wyre Steam Trawling Company and transferred to Fleetwood.

Her last report on 16 December 1948 stated she was making for shelter at Adalvik on the north-west coast of Iceland in severe weather. Tragically she disappeared with the loss of all twenty-one crew. The last message received from her was picked up by the Grimsby trawler *Lincoln City*, stating her intentions to seek shelter. Nothing more was heard from her.

On 15 November 1997 the Icelandic trawler *Helga* (RE 49) trawled up a ship's funnel in her net while fishing north-west of Halo, on the north-west coast of Iceland. The funnel was taken into Reykjavik; it was later identified as belonging to the *Goth*. The funnel has now been returned to Fleetwood where the relatives have preserved it as a memorial to the crew.

It was always terribly sad when a trawler went missing. Fishing communities were very close; everyone knew someone who had a relative or friend on a ship and everyone rallied round to help the families of the lost men.

By the time I was ten years old, I had already done five trips to sea with my dad on the *Comitatus*. One morning, just after breakfast, we were hauling our gear. Working off the west coast of the Shetland Islands in deep water we had 900 fathoms of warp out; we were fishing for hake and we still had about 75 fathoms of warp to heave up. It was then that we saw a large patch of the sea turning a blue-green colour with lots of bubbles appearing on the surface.

'Look at that,' shouted one of the crew. They knew what was coming next. Suddenly the net burst to the surface. It shot up out of the sea like a whale leaping out of the water – a huge haul of fish. The end of the net stuck out the water like a giant haystack. It stayed like that for a few minutes then slowly rolled over and spread out over the sea like a giant sausage.

The winch was slowed as we tried to heave the net slowly to the ship's side. Too much speed would've seen the net burst open and the fish lost, about 12 to 15 tons of it! Had the weather been bad we'd certainly have lost the lot. Fortunately the weather was fine but with long heavy Atlantic swells rolling along. We could see fish escaping; somewhere there was a hole in the net. After a struggle we got enough net on board to get a rope round it and heave all the fish down into the net.

I was very excited. I'd not seen so much fish in one haul but it looked as though we were going to lose most of it. The hole in the net was getting bigger by the minute and the sea was littered with fish floating away. Suddenly Dad jumped onto the deck, tied a rope around his waist, grabbed a mending needle and gave the end of the rope to a deckhand. 'Don't you dare let go of that,' he shouted, before jumping over the side onto the net and gingerly making his way along the back as I looked on in horror. I couldn't believe what he was doing. There was just about enough buoyancy in the fish to stop him sinking but I was petrified he was going to fall into the sea.

Once he'd reached the hole he did a quick repair and made his way back, more than once he nearly fell off. My heart was in my mouth and I was relieved when he was back on board.

When we'd got the fish on board the deck was absolutely full. We got about sixty baskets of hake – an excellent haul by any standard. The rest were coalfish which were no good. We gutted the coalfish to get the livers out for cod liver oil. In Fleetwood 'coalies' as we called them did not sell well so only the very best were saved for market. I thought my dad was a hero going out on that net, but after I'd had a few more years' experience I realised it had been a bloody stupid thing to do. I probably did things that were even more stupid! But that's a fisherman's life.

A couple of days later we were hauling the gear. I was sat on the casing watching the bosun working on the after trawl door. As he tried to put the chain on the door it swung inboard, trapping his arm. He screamed out in pain and fell to the deck. I was horrified at the amount of blood running down his arm onto the deck. I was shaking and felt sick, not knowing what to do. Dad came out of the wheelhouse and told me to get out of the way. I waited anxiously in the wheelhouse as they took him aft to assess the damage. Dad came and told me his arm was in a bad way and we'd be taking him to hospital in Oban.

One trip I was fooling around on the deck just passing time away with the deckhand on watch. He showed me how to mend nets and splice a rope. He was going to put some coal on the fire down in the forecastle so I went with him.

We went down below, he stoked the fire and we sat talking. He showed me how good he was at sticking his knife into the bulkhead. The rest of the crew was in their bunks trying to get some sleep. One of them got a bit pissed off with the noise the deckhand was making with his knife. A row followed and the deckhand was going to thump him. A voice from another bunk shouted, 'If you two don't make less noise, I'll get out and thump the pair of you, now f★★k off back on the deck.'

With that the two deckhands were up on deck squaring up to each other. Dad was on the bridge. On hearing the commotion he came out on deck. I thought he was going stop them, but no, instead he made them take all the deck boards up and put them out of the way then told them to get on with it, but to stop when he said. There was a bit of pushing and shoving, half a dozen blows followed, one had a cut lip and the other a black eye. Then Dad stepped in and stopped it. He made them shake hands and it was all over. The boards were put back and the ship returned to normal.

As Dad went back to the bridge he turned to me and said, 'That was nothing to do with you was it?'

'No Dad,' I replied trying to look innocent.

'If they're going to fight, let them, or they'll be at it the entire trip. If you let them fight they will become best friends,' he informed me.

And he was right, they did.

That trip we had a cat on board. Any stray animals we found at home always finished up at sea with Dad. This cat was as nutty as a fruitcake. It didn't matter how bad the weather was, he'd charge around the ship chasing seagulls he hadn't a hope in hell's chance of catching. He drove the crew mad by running along the ship's rail and straight up the rigging till he got to the top. Once at the top he would hang on and cry until one of the crew climbed up to get him down. He was threatened with a watery grave on more than one occasion.

On one trip a couple of deckhands brought some dye with them and threatened to dye the cat blue – Dad said just try, and that was enough to put them off. Instead they got buckets full of different coloured dyes and spent the rest of the day catching fulmars and dying them instead. There were red, blue and green fulmars flying around for the rest of the trip. Dad didn't think a lot of it and sacked the pair of them when we docked.

At home one winter we had severe frost and snow. At the back of our house was an outbuilding with a coalhouse and washhouse. One night I was told to go and get some coal for the fire. It was dark and snowing. I opened the coalhouse door. There were no lights inside; it was pitch black so I had to shovel the coal into the bucket by feel. I started poking about with my shovel looking for the coal. I kept poking something heavy, it didn't feel like coal. I put my hand out and touched the stuff. To my horror it was warm and it moved. I dropped the shovel and ran screaming into the house, face as white as a sheet. I was convinced I had found a body.

What made it worse was that about a year earlier a young girl had been murdered and her body had been thrown into a manhole in the road just round the corner from where we lived. The culprit had never been caught. I thought he'd come back, I was terrified. I ran into the house yelling to Mother that there was a body in the coalhouse. Dad was away at sea so she went next door to get the neighbour. He came round with a torch and we went out to the coalhouse. Standing well back while he went in, we could hear him talking to someone.

'Come on then, no need to be afraid, no one's going to hurt you now, we'll look after you,' we heard him say. We thought he would emerge with a body; instead he came out carrying a big black dog. Poor thing was in a terrible state, trembling, frightened, cold and very close to death. We got an old blanket out and laid him by the fire. We tried to feed him but he was too weak and feeble. Before we went to bed my mother built up the fire to keep him warm during the night. Dad was due home the next day and he'd know what to do.

Next morning the dog was still alive and looked a bit brighter. When Dad got home the first thing he did was give him a saucer of milk with brandy in it. He wasn't very good for a week or so but gradually got his strength back and made a full recovery. Dad took the dog, who we christened Blackie, to sea with him.

My next trip with Dad was early April (Easter holidays) that year. The weather was poor for the first four or five days. Wind was NW force 6 to 7, only just fishable. I loved the bad weather. I could stand on the bridge all day long just watching the seas rolling along, the bigger the better. There was something majestic and frightening about them. The crew used to tell me I was crazy. 'You won't think that when you've got to work in it,' they would say. But I'd just laugh.

A few days later the weather turned really bad. Within an hour the wind increased from force 6 to 7 to force 8 to 9 gusting to force 10. The crew was called upon to get the gear on board as quickly as possible. I was in the wheelhouse with Dad; he told me to keep on the port side out of the way. The mate and a deckhand went to the winch and started heaving the warps to bring the net up. The wind was on the port bow. When there was about 25 fathoms of warp to heave up the winch was stopped. Dad put the wheel over to starboard. The ship came slowly around. By this time the rest of the crew was on deck ready. When the ship was in position with the wind on the starboard quarter the engines were stopped. The order was given to heave on the winch.

The first thing out of the water was the forward trawl door; it was chained up and disconnected from the rest of the gear. The net was already on top of the water, the wind was screaming. The crew had to bend forward into the wind to stop them being blown over. Hailstones and heavy spray driven on by 50mph winds peppered their faces like icy needles. With eyes half closed, squinting to see, the crew went about their work without a word. No one spoke, every man knew exactly what to do, and did it. This is when teamwork comes into its own. And there's no better teamwork than a trawler crew getting the net aboard in such conditions.

When both trawler doors were safely in the gallows and unclipped from the warp the net was heaved to the ship's side. I was hanging on to stop myself falling over. Suddenly Dad dropped the bridge window! He shouted at the men: 'Look out for water! Get out of the way!'

Almost immediately a huge sea hit us. The sea crashed against the side of the ship, sending heavy sprays of water high into the air and through the bridge windows, filling the wheelhouse. I was soaked through. The ship was now laid over to starboard.

We were awash, the decks were flooded. I looked out of the window at the foredeck and it was full of water. Two of the crew had been swept from the starboard side onto the foredeck. They were struggling in the water, desperately trying to grab hold of something, anything, to regain their balance and get stood up. One was bleeding from his forehead. Suddenly I was scared.

Bloody hell, I'd seen bad weather before but never like this. I was petrified. Where were the rest of the crew?

Dad shouted at me to get down into the berth and out of the way. I didn't need telling twice. Blackie was down there. He lay on the mat, head on his paws. Half asleep he looked up at me as if wondering what all fuss was about. I got out of my wet clothes, dried myself off as best I could and sat on the bunk trying not to fall out.

There was a lot of commotion going on; men were shouting and the winch rattled as they struggled to get the gear aboard. I could hear Dad shouting orders from the bridge. After about an hour and a half I heard the ship's telegraph. The engine started up and the ship came alive. The old girl pitched and tossed as she struggled to get her head into the wind. Once we were up head to wind things settled down a bit though we were still rolling and tumbling about.

I heard the wheelhouse door open and voices on the bridge. I went up the steps to the wheelhouse but as I popped my head out the hatch I was told to go back down and stay there.

A bit later Dad came down followed by the deckhand, the one who'd cut his forehead.

'Are you OK Jim?' Dad asked me.

'Yes,' I told him.

He dragged the medicine chest out. The deckhand sat on the seat locker while my dad sorted him out. He had a bad cut just above his right eye which was closed and puffed up. Dad said it could've done with a couple of stitches but in view of the weather we wouldn't be able to get in to land until the weather eased off.

When he'd gone Dad told me not to go into the wheelhouse as the weather was now reaching hurricane force and it was too dangerous. I could hear the wind howling and the seas crashing against the front of the bridge. I had no wish to go anywhere! Little did I realise I'd be down here for the next four or five days. Food was stew and sandwiches and all the tea I could drink. My toilet was a bucket. The dog was well trained though. He'd walk the berth, looking up at the hatch. I'd shout to Dad and he'd take him up onto the wheelhouse veranda and tie a rope around him in case he fell off.

The time was endless as I'd nothing to do. I'd read everything that was down in the berth. I was bored stiff. I listened to the radio, although most of the time it was tuned in to other trawlers keeping in contact with each other.

The weather was relentless. We were rolling and tumbling about. I was bruised and aching all over from being knocked about and trying not to fall out of my bunk. I decided to get some sleep so I pulled the blanket over me; suddenly I awoke with a start. I must have dozed off. There was a loud crash as a huge sea hit us. The ship rolled violently over to starboard, it seemed she was going to roll right over. I was thrown against the side of the bunk. Even the dog jumped up and started barking. Now that did scare me! On the bridge I could hear shouting, the crashing of breaking glass, the splintering of wood as the wheelhouse windows caved in. The sea rushed into the bridge – it was full of water. There was only one way for the water to go: down the hatch and into the berth.

The water came down in such a huge mass it compressed the air in the berth. The pressure was so great it hurt my ears. I put my hands over them to try and stop it. As the flow of water eased the pressure became less, as did the pain. Now the berth was half full of water. The poor old dog was floundering about. He managed to scramble onto the seat locker and from there he made a great leap and landed in the bunk alongside me. I don't know who was more scared, the dog or me!

The ship was still laid over to starboard, I really thought she was about to sink. I made one leap from the bunk to the ladder. I had to get out of there. I missed the ladder and fell backward into the water. Panicking, I scrambling about until I got my grip. I climbed up, poking my head through the hatch.

'Are we sinking?' I shouted to Dad.

'No!' he replied. 'Get back down the berth, I'll be there in a bit.'

I managed to have a quick look round the bridge. It was total devastation. All the windows at the front had been smashed in. Wood and glass was everywhere and in a corner of the bridge sat one of the crew, his face covered in blood. I told Dad the berth was half full of water.

'Don't worry about that,' he said. 'We'll sort it in a bit.'

I noticed he'd cut his hand, it looked a right mess. I'd been convinced we were sinking but the ship was now more or less back on an even keel. The crew's injuries were not as serious as they looked. The damage to the bridge, however, *was* quite serious. The windows had been completely taken out and the plates at the front of the bridge had been buckled. There was damage around the deck and we'd lost a few deck boards. We'd been extremely fortunate really. The first job was to secure the bridge by nailing boards and canvas across the gaps where the windows had been. The crew then set about bailing out the berth and getting things dried out.

Twenty-four hours later we were steaming towards Cape Wrath, homeward bound. The sun was shining and it was a glorious day. The weather was down to a force 5 or 6, although there was still a lot of swell. It's hard to imagine that a day earlier we had been in weather so violent I'd wondered if we were ever going to get out of it. On the way home Dad asked if I'd be back to sea again after this.

'Just try and stop me,' I said.

'Then you are crazier than I thought,' he replied.

But I think he was pleased with my reply.

That was my last trip on the *Comitatus* as she was sold to a Grimsby firm that year. I felt quite sad as over the years I'd got to know her like an old friend. I spent many happy hours on her and got to know a lot of good shipmates.

I did one more pleasure trip with my dad in the summer of 1952 on the trawler *James Lay*, but it was not the same. The *James Lay* stopped for a fit-out and Dad went mate on the *Reptonion* to Iceland. That ship changed all our lives.

My dad was so taken with fishing at Iceland that on returning to port he asked the firm if he could go Icelandic fishing permanently. They refused. He was one of their top hake skippers. He was told to go back in the *James Lay* or they wouldn't give him another ship. So he left Fleetwood, joined Northern Trawlers in Grimsby and went mate on the trawler *Northern Isle*.

CHAPTER 4

THE GREAT STORM OF 1953

January was the month of the great storm. It started as a gale 400 miles out in the Atlantic but soon became a major storm heading for the west coast of Scotland.

On 30 January the Fleetwood trawler *Michael Griffith* left port after repairs to her boiler. In the early hours of the following morning she was 6 miles south of Barra Head on the west coast of Scotland in the full force of the storm, the wind from the north-west gusting up to 100mph with 30–40ft seas.

A distress message was sent out saying, 'We have no steam. We are full of water. We are helpless! Come and help us?' Nothing more was heard from her. This was a truly desperate situation. With no steam there'd be no power for the engines to keep the head up into the wind. This is vital to survive in such atrocious weather conditions – without it the ship would be wallowing broadside in heavy seas. She sank with the loss of all thirteen crew, the first casualty of the great storm.

The Grimsby trawler *Sheldon* sailed 30 January from Kirkwall in the Orkneys bound for the Faroe Islands. The following day she was presumed sunk in storm-force winds 60 miles north-west of Dennis Head with the loss of thirteen men.

That same day the mail boat *Princess Victoria* radioed Port Patrick stating they'd left Stranraer for the 35-mile voyage across the North Channel to Larne. The mail boats always sailed regardless of the weather. Within half an hour they were in the teeth of the storm. A massive sea flooded the car deck causing the vessel to list. At first this was not a problem, but as the day wore on the list grew steadily worse. The Port Patrick radio received a triple-X message, which is one below an SOS distress call, saying she was listing badly and required the assistance of a tug.

When they were 4 miles off Corsewall Point, the radio operator, David Broadfoot, sent out a distress call requesting immediate assistance. With winds reaching up to 120mph and 50–60ft-high seas, the Port Patrick lifeboat was launched. She headed off to the position of the stricken vessel but could find no trace of her. Port Patrick radio asked David to keep sending his Morse signals so that a bearing could be taken of the ship's position.

The *Princess Victoria* was now laid on her side and the order given for passengers and crew to don their lifejackets. The ship's siren sounded and the order to abandon ship followed. David was still sending out messages. Within minutes of the order to abandon ship the *Princess Victoria* turned over. No more signals were received.

None of the survivors were women or children. One survivor said he saw a lifeboat full of women and children that appeared to be well clear but the next thing a sea rolled along and took the lifeboat back into the side of the ship. The lifeboat capsized, throwing the women and children into the water. Of the 176 passengers and crew only forty-three people survived, due to the heroic efforts of the Donaghadee lifeboat. David Broadfoot was awarded the George Cross for his bravery and devotion to duty.

Two hundred and thirty people lost their lives at sea during the storm, and nine ships were lost.

The storm turned south, down the east coast of England. Heavy rain had been falling for several days and the rivers were already swollen. The winds reached speeds of 125mph. Combined with high spring tides this caused a sea surge, pushing a wall of water across the North Sea.

In the early hours of Saturday morning the storm hit the Lincolnshire coast, destroying sea defences and houses. The storm devastated places such as Mablethorpe, Kings Lynn, Dersingham and Heacham. Then it hit Great Yarmouth and Southwould.

In all 300 people died and 24,000 homes were destroyed or damaged.

The storm devastated Holland, drowning 1,836 people, flooding 400,000 acres of land and damaging 50,000 buildings, 9,000 of which were totally destroyed.

I was still at school but I followed the reports of the tragedy as it unfolded in the local newspaper and on television. It was grim reading and it left a lot of people devastated. Yet as tragic as it was it was never going to put me off going to sea.

CHAPTER 5

ST PHILIP, 1953

I left school in the April after the great storm. I was ready to go Down Dock and get a ship but my mother said I wasn't going anywhere till Dad got home. He was at sea on the *Northern Isles* and it'd be two weeks before he was back.

When he came home he told me not to be in too much of a hurry but to have a few weeks at home and get the bad weather out of the way in view of the *Michael Griffith* disaster. One of her crew was a young lad of sixteen and it upset my mother a bit.

I spent the next six weeks waiting to get started. Finally Dad took me down to the Boston Deep Sea Fishing Company's office, which surprised me a bit as I thought it would've been the Dinas Steam Fishing Company, seeing that he'd been with them for so long, though the Boston Deep Sea was one of the best firms in Fleetwood at the time.

On arriving at Boston's I was taken upstairs to the gaffer's office. The gaffer, Arthur Lewis, was a successful trawler skipper himself. He told me I'd be going as brassy on the *St Phillip* with skipper Tom Hodgkinson, also known as Soccer. About five minutes later he arrived at the office and I was introduced to him. He was a short stocky fellow and looked to be about thirty. Arthur told him to look after me and make a good 'decky' out of me. 'I will, but if he's cheeky he'll get a kick up the arse,' was his response. I found out later he meant what he said. The *St Philip* was having a few repairs done so it'd be two or three days before we sailed. My dad and Arthur Lewis shook hands as we left.

We went downstairs to the runner's office. The runner was Rupert Robins; he recruited the crews for the Boston ships. Dad asked him to sort out some sea gear for me then left. Rupert had been with Boston's for a long while and was well liked and respected by the crews. He told me what a good firm Boston's was to be in. Then he gave me a chit and sent me across the road to the pool office to sign on the *St Philip*.

The pool office was at the dock gates. As I entered the place it was full of cigarette smoke and men hanging about for any jobs to be had that day. It was noisy as they all chatted to one another. The pool office was used by all the trawler firms in Fleetwood; as ships came in dock the logbooks went to the office. When the firm gave us a ship we were given a chit to take to the pool stating what ship we were to sign on. I went in there not sure who I was to give my chit to. Then one of the men said, 'You signing on lad?'

'Yes.'

'Then you want to be over there.'

'Thanks.'

I went across to the counter where they had all the logbooks and handed in my chit. The man took it and said, 'You Billy Greene's son?'

'Yes.'

'You want to be a skipper like your dad?'

'Yes.'

'Well if you turn out like him, you'll be a good 'un. Sailed with him in the *Hayburn Wyke* – bloody good skipper. There you go lad, sign here.'

He pushed the logbook towards me and I signed on for the *St Philip*. I walked out of that office buzzing. I was so excited I could hardly believe it. I was now officially a Fleetwood fisherman and very proud of it.

I went back to Boston's. Rupert told me the ship would be sailing on Thursday. He then took me to stores where I was given new gear: an oil frock, a pair of thigh boots, a sou'wester, a couple of mufflers, some woolly boot stockings, a new jersey and a pair of heavy trousers, but no gloves. In those days Fleetwood fishermen did not wear gloves. All the gear except heavy sea gear went into my new kit bag.

'Take the kit bag home with you,' said Rupert. 'All the heavy gear will be on board for when you sail.'

Rupert took me to the cashier's office and told them I'd just signed on the *St Philip*. The cashier gave me a half crown sub, which believe it or not is 25p. I left and Rupert warned me not to be late for sailing. I knew there was no chance of that. I was so proud. I walked all the way home with my kitbag on my back, it was nearly as big as I was, and half a crown in my pocket. I walked into our house with the biggest smile you can imagine.

On the Thursday I arrived Down Dock in a taxi paid for by the firm. The firm always sent a car for the crew going to sea but you had to find your own way home when you docked. There were four people in the taxi; nobody spoke for the entire journey. The *St Philip* was laid broadside to the quay. My first impression was what a rust bucket! She was badly in need of some paint; the funnel was so badly streaked white with salt-water stains and seagull drop-pings that the colours were hardly distinguishable.

The crew where already there, waiting to sail. I looked around hoping to see a familiar face. No such luck. Rupert, the ship's husband, came over to me. 'The crew is all here Jim, better get on board, give me your kitbag and I will hand it down for you.' I climbed onto the ship's bow, Rupert handed me my bag, and I went down the ladder onto an unfamiliar deck. I stood looking around wondering what to do next. Someone said, 'Go and give them a hand aft lad.'

Suddenly I felt very vulnerable and alone. No Dad to look after me now – I was on my own. Was this really a good idea or a big mistake? Was I doing the right thing? Doubts were flooding my mind.

The bow ropes were cast off. As the stern rope was taking the weight the skipper spoke his first words to me, or should I say yelled. 'You,' he bellowed. 'Get out of the f***ing way; if that rope parts it will take you're f***ing head off!'

I've never moved so fast in all my life – nobody had ever sworn at me like that before. Suddenly I felt like bursting into tears. I desperately wanted to get off. I looked up at the quayside; it was too far off to make a jump for it. There was no way off now, I was on my own.

I stood quietly on the deck as the ship passed through the lock pits. I suddenly felt very disheartened, not a good start to my fishing career: I'd only been aboard ten minutes, had a bollocking off the skipper and no one was speaking to me. Fortunately it didn't last long. Once we were through the lock gate and into the River Wyre everything changed. The crew became chatty and friendly as they set about clearing the decks. There were nets, wooden bobbins and wires of all descriptions. There were the cook's provisions to go away, the meat had to be stowed and packed in ice down the fish room. The anchor chain had to be cemented down to stop water getting into the chain locker. The decks were littered with coal dust and had to be swilled down. Anything loose was lashed down, so it didn't get washed away.

The weather was poor that day. By the time we got to the Wyre Light we were heading into a full north-west storm.

The deckhands always slept down the forecastle. Situated in the ship's bow it's not the best place to be in a gale. The entrance was through a hood way and down some ladders. The forecastle is a V-shaped space and at the fore end there is a bogey stove. There are bunks for six men: on each side there are three bunks one on top of the other. I was told that my bunk was at the bottom port side.

The bogey stove had been lit, filling the place with coal smoke and burning wood. The forecastle smelt of tar mixed with the smell of fishy oilskins; it was musty and damp. Nailed at the bottom of the bunks on each side was a plank of wood that sufficed as a seat.

The ship's bow was rising and falling as she rode the seas. I was hanging on, not daring to let go. When I did move, it was a case of let go and make a dash to the next nearest thing I could grab hold of. As the weather got worse the ship's movements became erratic and violent as she crashed into the seas. My stomach was going up and down with every rise and fall of the ship. I didn't feel well, so lay down in my bunk.

We were steaming head into the seas. The ship's bow climbed up the sea; it seemed to climb so slowly, staggering and lurching up to the crest of the sea, hanging briefly in mid-air before crashing down with great force into the trough of the next. When it reached the bottom, the bow crashed into a wall of water – it was like hitting a brick wall. You could hear the sea crashing over the whaleback (bow) and cascading down the decks. Heavy spray lashed the ship and she shook and shuddered as the water travelled aft. Once again you could feel her lifting to the next sea then it all started again. Up and down went the bow and my stomach going with it.

I was hanging onto the side of my bunk for grim death. I was sweating and feeling clammy, the fire was in full flow and the forecastle was as hot as hell, and with the heat it stank even worse, which didn't help. One minute I was being hurled upwards, then when the ship plunged down my bunk fell away from underneath me. I was lifted into the air, banging my head on the bunk above me which was only about a couple of feet away. Then as the ship hit the bottom of the trough I was unceremoniously thrown into a heap at the bottom of my bunk.

The feeling in my stomach was terrible. One minute it felt like a knot in the pit of my gut, the next thing it felt like it was in my mouth. I felt greener and queasier with every second. Suddenly it was on its way up; my guts knotted up, it started deep inside and made a rush for my throat. I virtually fell out of my bunk and shot up the ladder. I just managed to get to the top and stick my head out of the hood way. I was violently sick. I couldn't believe it. I was actually being seasick. I'd done fifteen trips on a trawler with my dad in all sorts of gales and violent storms and had never been seasick. Yet here I was on my first trip, starting my life at sea, less than two hours out of the dock, and I was as sick as a dog. It lasted for about two days before I got over it. I really thought I was going to die.

We had started fishing before I had properly recovered. I used to laugh and take the mickey out of my mates who were seasick. If I knew how they'd been feeling at the time I don't think I would've been laughing so hard. You just can't believe how bad it feels, it's the only time in your life when you really wish you were dead. I was like that every trip for the next six months.

The following day the weather was a lot better, the wind had eased and the sea was calm. All crew was called out at 06.00, including me. As I was now part of the crew there'd be no getting up at ten o'clock and having Albert boil me an egg in the soup pan.

The mate was Albert Head. He came from Lowestoft and a right old seadog he was. He was great, he used to look after me and made sure I didn't do anything I shouldn't. He was one of the hardest men I ever met. Sometimes he would get onto the skipper for being too hard on me, but the skipper would tell him to 'F★★k off and mind your own business'.

Fleetwood crews never wore any kind of gloves working on deck, so their hands became hard. The twine we used for the nets was made out of Manila and sisal; it was very coarse with fibres sticking out of it. When I was filling the needles the fibres would stick into my hands like sharpened pins; they ripped your hands to pieces until you hardened to it. All the nets were dipped into a tarry solution to preserve them. When I filled the needles I had blocks of tallow to rub on them, this made them easier push through the net.

The skin on Albert's hands was like leather. I remember once he had a poisoned hand and he decided to put a poultice on it; this stuff was in a tin and had to be heated before use. Albert put it in a pan of water on the stove and went back on the deck. Twenty minutes later he came back off the deck. The pan of water had been boiling away for twenty minutes. He lifted the tin out of the water with his bare hands, put it onto the draining board and spread the poultice on a piece of lint; he then slapped it onto his hand. 'Phew! That was hot,' he declared. Any normal person would have screamed out in pain with a scalded hand.

Albert was also a survivor from the Fleetwood trawler the *Dhoon* when she was wrecked off the Icelandic coast in 1947. He never said much about it, but three of the crew lost their lives, and the rest were rescued from the shore by breeches buoy.

I don't remember very much about the crew. It's strange, on some ships you can remember most of the crew but on others you can't remember any of them. In my life as a fisherman I got to know many men, they came and went, some stayed one or two trips and you never saw them again. The regular crews were local people you already knew and got to know very well; they stayed with the same ships and firms most of their lives.

I didn't have breakfast that morning, as I was still feeling a bit seasick. At 06.30 we all went on the deck. The rest of the day was spent getting the nets ready. We had two lots of fishing gear, one on the port side and the other on the starboard side; these had to be made ready for fishing. My job was to fill the needles used for mending the nets. I could do this but was a bit slow. I'd done it for the crew on the *Comitatus*, but apart from this I didn't know much at all and felt a bit useless. By twelve o'clock I was feeling a bit better.

I went down the cabin to have some dinner; the cabin was situated underneath the ship's galley on the after end of the ship. As you went down the ladder, on the port side was a small berth; this belonged to the chief engineer. On the starboard side of the cabin was the mate's berth; between the two was a coal-burning stove. At the after end of the cabin was a table with seat lockers all round it. On each side, above the seat lockers, there were three bunks for the second engineer, two firemen, the boson and the cook. All the deckhands and the brassy slept down below in the forecastle. This was the same in most ships; it was all a standard design. The skipper had his own berth under the wheelhouse.

The cabin was hot, smelly and stuffy. I had soup and bread. By the time I'd eaten it I wasn't feeling too good to say the least. I left the dinner table pretty smartish and proceeded to throw up my dinner over the after end of the ship.

We were all working on the foredeck when the skipper shouted out of the window to me to get him a pot of tea. As I went aft one of the deckhands said to me, 'When you speak to the skipper call him Tom; he hates being called skipper by his crew.' I left the galley with a full pot of tea – by the time I got to the wheelhouse it was half-empty. 'I can see you need some practice on carrying pots of tea to the bridge' the skipper said. 'Half a bloody mug of tea's no good to me.' 'OK Tom,' I replied. He hit the roof. '*Tom*?!' he bellowed. 'It's f★★king skipper to you. Nobody on this ship calls me Tom, let alone a f★★king snotty-nosed little sod like you. Now get off my bridge and don't come back till I tell you.' I was out that wheelhouse like a shot. When I got back on deck I was shaking. Albert, the mate, asked me, 'What was all that about?' When I told him, he turned to the decky. 'One more stunt like that and you'll have me to deal with.' The deckhand didn't dare say a word; you didn't argue with Albert. He was built like a grizzly bear and didn't take any backchat, but he was a fair man, and a very good mate.

Later that afternoon the skipper called me back to the wheelhouse; nothing was mentioned about the earlier incident, he was quite friendly. 'I'll help you all I can,' he said. 'But you have got to knuckle down, and if you're any good we'll make a good deckhand out of you.' He also told me that this trip he wasn't expecting anything from me. 'Use this trip to get to know the crew and don't take any notice of anyone except the mate,' he explained. 'Next trip we will start for real.'

The rest of the trip was uneventful. I filled the needles for mending the nets, helped the cook and generally did my best.

When I got home after that first trip everybody made a fuss of me and asked how I'd got on. Next day I went down to the office to get my pay which was about £3. I was also given a parcel of fish known as a fry – no guessing what was for tea that night. After we all went to Blackpool and had a good night out. As the night wore on the lack of sleep was catching up on me, I was so tired. We sailed the following day after thirty-six hours in dock – I didn't feel like I'd been home at all!

The next day I arrived at the ship ready to sail and found her laid alongside the quay all smartly painted up, her hull shiny black with a white strip all the way round. Her handrails where all painted white and the funnel was painted black with two narrow red bands two thirds of the way up. I looked at her with pride.

This trip we steamed off to St Kilda and fished in the deep water. When I'd got over my seasickness I was ready and raring to go.

The *St Philip* was a good sea boat, which was just as well, as we fished in a lot of bad weather. Once when we were fishing in bad weather the skipper got two canvas floats which he filled with oil. As we were laid broadside to the sea, he lashed each float to the ship's rail and threw them overboard. The idea was that the oil would slowly seep out, spread over the sea and calm it down a bit. As the saying goes, 'Oil on troubled waters'. I must admit it did work, and it did smooth out the sea a bit, although it didn't make the seas any smaller. One problem was the oil washed back onto the deck and we could hardly stand up for the next two or three days it was so slippery. That was the last time he tried that.

We worked our way round from St Kilda to the Butt of Lewis. The weather was lovely with a calm sea. We hauled our gear up, and the skipper told the mate to get the net on board and lash it down. When the fish was out of the way down the fish room we battened all the hatches down. The skipper said the forecast was for storm force 10 imminent so we steamed into the shelter of the Butt of Lewis. When we arrived the sun was still shining and the sea lovely and calm. We all went below to get some sleep.

The following morning we had breakfast and spent the rest of the morning getting bits of fishing gear ready to use. The weather was still sunny and calm, so most of the afternoon we spent sunbathing and generally relaxing. At teatime there was still no sign of this storm. The skipper said, 'F★★k this! We'll steam off to the deep water and try and get a night's fishing in before the storm arrives.' We had been laid for thirty-six hours in glorious sunshine waiting for a storm that never came. 'Next time we get a storm warning we'll carry on fishing till it does arrive,' he added.

The system Fleetwood trawlers worked for young lads starting out in the industry was quite good. You'd start off as a brassy and the pay was a quarter of that of a deckhand. After that it was up to the skipper. If I worked hard, learned how to mend and generally did well, then he would tell the office to upgrade me to half-decky, then I'd get half a deckhand's money. With further progress I'd get three quarters of a deckhand's pay, after this you'd become a full deckhand. If you were a brassy who didn't show any interest in the job or learn anything then after three months you'd be sacked. I was keen to learn and wanted to do well so I did my best.

One morning skipper said it was time for me to learn to mend. We worked two gears, one on the port side and one on the starboard side. The starboard side gear was in use but the port side net was laid on the deck. It was the mate's watch so the skipper took me on deck to teach me how to mend. The first thing he showed me was how to mend was a 'halfer'. This is the easiest hole to mend on a net; it's when a single strand in the mesh is broken. He showed me once then showed me again.

'There,' he said. 'Do you think you can do that?'

'Yes,' I said. 'That looks easy.'

He handed me the needle. 'Your turn then,' he said.

So I tried. When you start mending for the first time, it's very awkward and you're all fingers and thumbs.

'You can't do it can you?' he said.

'No,' I replied.

'Give me the needle and I'll show you again,' he said.

I passed the needle back to him; he took it, then out of the blue hit me right across the back of the hand with it. I wasn't expecting that and it brought tears to my eyes. What I thought about him at that moment isn't repeatable!

'In future,' he said, 'you don't say you can do it when you can't.'

After that I'd pass the needle back to him and stand well out of his way.

I often got a slap with a needle or the back of his hand. One day he was showing me how to cut out a hole in the net ready for mending. He was holding the net up and I was slashing away with my knife. I was doing what he told me but once or twice got a bit close to him with the knife.

'Watch you don't cut me with that bloody knife!' he barked. Well that was the worst thing he could've said, because next thing I'd accidentally slashed his hand.

'You little bastard!'

He made a grab for me but missed. I shot straight up into the wheelhouse. I knew that if I got to the mate before the skipper got to me the mate would calm him down. When the skipper *had* calmed down he was back to normal, besides it was only a scratch on his hand. The way he'd reacted I thought I'd cut at least one finger off. Although he was hard on me he was OK and I was never really frightened of him; he was a bloody good teacher and I owe him a lot. By the time I left him the following summer I was pretty good at mending nets and all the other jobs deckhands do.

The *St Philip* was a coal-burning trawler and she carried two engineers and two firemen. The coal was situated in the bunker amidships. We also had two wing bunkers. Through the main bunker was a tunnel leading into the after fish room.

When we left the dock the after end of the fish room was full of coal and it was boarded off from the fore part of the fish room; the firemen had to take the coal out of the fish room first. When there was just enough coal left in the fish room to fill the tunnel, the door to the engine room was closed. The firemen then had to shovel all the coal that was left into the tunnel ready to start the next trip.

The fireman had an awful job. He spent his watch in the stoke hole shovelling coal into the furnaces to keep the fires going, pulling the ashes onto the deck and dumping them over the side. The stoke hole was a narrow strip between the boiler face and the coalbunkers. It was as hot as hell in there and I hated it! I never went down there unless I had to. It was a dangerous place to be, especially in bad weather when the ship was rolling about; there was always the danger of coal coming down on top of you. I was sent down there on a couple of occasions to help the fireman; I sweated buckets climbing up and throwing lumps of coal down. I really hated it.

The coal had to be used as economically as possible to make it last the trip. When the fishing was good we'd pull the gear on board in the dark and lay to in daylight hours to preserve coal. We'd get the gear ready for the next night's work then grabbed some sleep; we started fishing again after tea. This trip the fishing had been good, so we needed to get the after fish room emptied.

The skipper called me to the bridge and sent me to give the fireman a hand to fill the tunnel and empty the after fish room. The fireman was a Welshman; he wasn't very well liked as he was a trouble maker. The chief engineer sorted us out with three carbide lamps. Once down the fish room one of the carbide lamps went out and the fireman started messing about with it. I wasn't very impressed as they were dangerous and could easily spark an explosion.

Years earlier when the *Comitatus* was named the *Betty Johnson* there'd been an explosion caused by carbide. The engineers were handing a drum of carbide down the engine room and in the process they'd dropped it. When it hit the engine room plates it burst open. The men didn't notice that some carbide had dropped into the water in the bilges where it began to give off acetylene gas. The gas built up in the engine room. When it reached the boiler room, and the fires, it sent an explosion ripping through the ship. It resulted in two of the crew being killed and three seriously injured. It's not a substance to play about with.

Once we were down the fish room the crew battened down the hatches. We were there till all the coal had been shovelled into the tunnel, which took us about four hours. We had to shovel the coal into fish baskets and drag them along to the end of the tunnel, then tip them out. We had to make sure the coal was stacked right up to the top of the tunnel otherwise we'd not get all the coal out the fish room. It was hard, backbreaking work as the tunnel was not high enough to stand up in. All the time I was crawling up and down the tunnel dragging baskets of coal and stacking it into the tunnel; the fireman's excuse for not helping was he had a bad back.

We eventually filled the tunnel and emptied the fish room, put a wooden plug over the tunnel and made it secure. We hammered on the fish room stanchions to attract the crew; when they finally heard us they came and let us out. It was good to see daylight again. I was absolutely knackered and as black as a chimneysweep. The after fish room was washed out, the bilges cleared of any coal dust, so we didn't get a blockage of the fish room pumps. The fish room was then shipped up ready for use.

The skipper sent me to get washed and turn in for a few hours' sleep. After I'd had a wash I went to the galley and had a cup of tea with Albert.

'Well Jim, now you can say you've worked with a murderer.'

I nearly choked on my tea! It transpired this Welshman had committed a robbery in Wales and whilst trying to escape he'd hit somebody over the head with a lead pipe and the man had died. He'd only just been released from prison after serving his time. It's a good job I hadn't known this earlier because there is no way I would've ever gone down into that fish room with him!

We were fishing at St Kilda. Fishing had been good and we were getting a reasonable night's work. We'd been laid to during daylight hours and the first haul was after tea. We had two bags of small haddock, and it carried on like that throughout the night. Gutting haddock without gloves is no joke. Haddock guts are usually full of small pieces of crushed shell, and while gutting these bits of shell gets between your fingers and chafes away the skin until your fingers are bleeding and sore. This was nicknamed haddock rash. They also get round your wrists and chafe the skin causing saltwater boils, which are very painful. Some of the men I sailed with suffered very badly with them. I sailed with one bosun who used to scrub them with a dry scrubbing brush to try to get rid of them. That must have hurt like hell! Another trick I saw was when the boils had come to a head and were ready for bursting, the crewman would get an empty milk bottle, hold it over the spout of a boiling kettle, and when the bottle was full of steam place the neck of the bottle over the boil and press down on it. As the steam evaporated it caused a vacuum in the bottle and sucked out the pus. I had it done to me once and the pain was unbearable, it was less painful to keep the boil.

We finished gutting the haddock about mid-morning. We squared up the deck and had dinner then went below for a few hours' sleep. When I rolled out my fingers were stuck together with congealed blood and boy were they sore. I had to get a bucket of warm water from the galley, soak my hands, and then gently prize my fingers apart, a painful operation. The following night we had more good fishing. By the end of the second night my hands were in a terrible state. I wasn't crying but by God I was very close to it. The rest of the crew's hands weren't too bad as they were used to it. It took a few days before my hands healed up properly. Working in salt water tends to slow the healing.

The next night the first two hauls only produced a few baskets of fish so we spent the rest of the night steaming to another fishing ground. I found out later the reason there had been no haddock that night was one of the deck-hands had cut a small hole in the cod end (the part of the net that holds the fish in) so all the fish escaped. When the skipper found out he informed the office and the man was barred for life from the firm.

I thought the haddock gutting was bad but even worse was the night we ran into a shoal of spur dog off Barra Head. Dogfish are like small sharks, have a spike running alongside the dorsal fin and have skin like sandpaper. We caught 400 kits of these dogs in eighteen hours. They all had to be gutted and they rip your hands to shreds. They chafe away the back of your fingers on your right hand. When you gut them you're constantly running your fingers down the belly and they finish up red raw and bleeding. On your left hand it's your finger ends and the palm of your hand that gets it. I still have small scars on my knuckles to this day through all the chafing. I must admit I did silently cry that night. Fortunately we were heading home after that. It was a lot of hard work as dogfish fetch very low prices on the market. When I got home my mother just couldn't believe the state my hands were in.

After I'd done a few months in the *St Philip*, I was getting on very well and I'd already got to three-quarter-deckhand's pay. But winter was drawing close and the weather was getting worse. One morning at breakfast the skipper told me I'd not be going on deck. The weather was a bit scruffy and I would be spending the day on the bridge with him. He was going to teach me how to steer the vessel and learn the compass. When I went to the wheelhouse the skipper was already there, he'd relieved the watch and sent them for breakfast.

'Right what do you know about the compass?'

'Nothing,' was my reply.

'Were you any good at school?' he asked.

'Average,' I said.

'Well you'll have to be better than average this morning because if you don't know this compass by twelve o'clock you get no dinner.'

Bloody hell, I thought, that's my dinner gone for today, I'll be starving by teatime. It was 06.30 and I had to learn thirty-two points of the compass by twelve o'clock.

'Right, we start at North and work through the compass points till we get back to North,' he said.

'OK,' I said.

'We'll start with the first four points that will be North, North by East, North North East, North East by North.'

'Right,' I said.

'When you've learnt them, we'll then learn the next four points and then put both lots together and so on. You'd better get on with it. Let me know when you've done the first four.'

'Right,' I said.

It was one of the longest, most boring mornings I've ever spent. I learnt the first four points of the compass. Good, I thought. By the time I'd learnt the second four, I had forgotten the first four. But we plodded on, saying them over and over, again and again and again. By the time dinnertime came I had learnt the compass in a fashion. I could recite it, and although I had to be prompted on several occasions, I could manage it. At twelve o'clock the mate came to relieve the skipper.

'Do you know the compass yet, Jim?' Albert asked.

'Just about,' I said.

'But not good enough,' said the skipper. 'He's not getting any dinner until he knows it properly.'

Albert asked me to repeat the compass. I repeated the compass, and for the first time I repeated it almost perfectly.

'That was OK, Jim,' said Albert, then he said to the skipper, 'I think he's done a good enough job to have earned his dinner.'

'OK, go and get your dinner,' he said.

'Great,' I said. I felt I'd done well; I was quite pleased with myself to be honest.

After dinner we were back in the wheelhouse. The skipper asked me to repeat the compass again. Big mistake: I'd had dinner and a good half hour since I'd last repeated it and guess what, I'd forgotten half of it. The skipper called me all the thick bastards under the sun. 'This afternoon find a quiet place somewhere and keep going over that compass till you can say it off by heart.' I did, and come teatime I could say it without any mistakes whatsoever.

On the bridge next morning the skipper asked me to repeat the compass again, and I said it perfectly.

'That was good; do you know the compass now?'

'Yes,' I said, feeling proud of myself.

'Good, now say it backwards.'

'I can't,' I said, somewhat taken aback by the request.

'Then you don't know the compass,' he snapped.

Off we went again, and I had to learn it backwards. By the end of the trip I could repeat it both ways. It was only when he taught me to steer that I realised the importance of knowing it both ways.

A couple of days later and the weather was too bad for fishing. We were steaming slowly head into wind, known as dodging. In the wheelhouse the skipper turned to the mate.

'Let Jim take over the wheel and have a go steering.'

'Do you think he'll be all right?' replied Albert.

'Yeah, let him have a go.'

I took the steering wheel. I'd steered the ship before but only in fine weather. This was a new experience for me.

'Now hang on to that wheel and don't let go,' instructed Albert. 'If that wheel kicks and you don't hang on it'll be ripped out of your hands.'

In the 1950s there wasn't any hydraulic steering gear unless you were fortunate enough to be in one of the new modern diesel trawlers just coming out from the yards. Our steering worked on a chain race. The chain ran from the rudder arm on the port side aft, along the deck up through the wheelhouse, back down the starboard side of the wheelhouse, along the starboard side deck and aft to the rudder arm. Whilst I was at the wheel it pulled and kicked for most of the time. But it was when we were going over the top of a sea and the rudder came out of the water that I really struggled. If the rudder was hit by a sea while out of the water the weight of the sea would try and push the rudder hard over, this is when I really had to hang on or it'd rip the wheel out of my hands and spin out of control – the danger being that the chain could snap and leave us with no steering or worse the rudder would be forced over so hard there'd be a danger of smashing the rudder post. This had happened on the *Hayburn Wyke*, so I needed to be on my toes.

The following trip we were working the deep water off North Rona when I was called to the wheelhouse. 'Here,' the skipper said. 'Someone wants to speak to you.' He handed me the radiotelephone. Who on earth would want to speak to me I wondered. I was surprised and delighted when I heard Dad's voice. He was mate on the *Northern Isle*. They'd been scattering a skipper's ashes at sea, were 10 miles away and steaming towards us. I was very excited as I'd not seen a deepwater trawler from Grimsby or Hull before. When they arrived I was surprised at the size of the ship.

They came alongside of us and passed a couple of bottles of whisky over. I'd not seen my dad for a few months; when he was in dock I'd be at sea or vice versa, so we didn't see much of each other. They were on their way to Iceland to fish; their last trip there had landed 2,800 10-stone kits of fish, mostly cod. I couldn't imagine that amount of fish, if we landed 500 kits it was a good trip.

A few days later we went into Castle Bay on the island of Barra. Our skipper was telling another skipper what a good brassy I was and how well I could read the compass, so he asked me to read the compass point our ship was heading and I made a right cock-up of it. I said something stupid like south-west by east. The skipper wasn't very impressed and smacked me one. 'Go easy Tom, there's no need to do that to the lad,' said the other skipper. But skipper said I deserved it for letting him down. I felt pretty rotten about it, though I don't think the skipper had meant to hit me so hard.

The next day we steamed to the deep water at St Kilda. The weather was poor. We hauled after breakfast, and when the net came to the surface it was in ribbons. There was tattered net all over the place. The cod end was floating on top of the water; it was full of fish, mostly hake. We were heaving very slowly trying not to damage the net any more, but we were fighting a losing battle, the net was gradually falling apart. Then the worst happened and the net completely fell in two. Now the bag of hake was floating, away much to the dismay of the skipper. He screamed from the bridge window, 'Get that gear aboard as fast as you can, and get a grapple ready!' Bloody hell, I thought, he's going to go after the bag of fish. As soon as the gear was on board, he rang the engine room for full speed. He swung the ship's head round till he spotted the fish floating in the distance, then straightened the ship on the same course and headed for it. The crew thought it was hilarious and that he'd no chance of ever getting it on board, especially as the weather was not in our favour. Credit where it's due, he manoeuvred the ship right alongside the bag of fish, we threw the grappling at it and got a good hold on it first go. We got a rope around the net and heaved it on board. We lost most of the fish but I gave the skipper 100 per cent for trying.

CHAPTER 6

HILDINA, 1953

The following trip we sailed at the end of November. The weather was bad so we steamed through the Sounds of Mull to the Butt of Lewis. Quite a few trawlers were laid under the Butt sheltering from the weather.

It was Monday 30 November 1953. We laid most of that day; the weather wasn't getting any better. At 18.00 the skipper decided we'd go into Stornoway harbour and tie up for the night. The next morning around 09.00 the skipper broke some bad news, the Fleetwood trawler *Hildina* had been overwhelmed by heavy seas and had foundered about 50 miles north-west of the Sule Skerry. Her sister ship, the *Velia*, was heading for her position. There was no news as to the safety or whereabouts of the crew.

This was the second major disaster for Fleetwood this year, the first one being the *Michael Griffiths* lost in the great storm. The *Hildina* was built for J. Marr & Sons. At 296 tons she was one of the first of six diesel-driven trawlers built the year before. She was the first word when it came to luxury accommodation and development.

Earlier that morning, Tuesday 1 December, the *Hildina's* net had become fast on the seabed. What started as a normal day's fishing fast became a tragedy. With her nets fast she took a succession of massive seas. The skipper, George Clarkson, gave the order to reverse the ship. It turned out to be his last command. Within minutes the ship was awash and heeling over badly to starboard. The crew tried desperately to chop away the gear, but they couldn't reach it, they were up to their chests in water. The rest of the crew who were aft rushed on deck, tore off their heavy sea gear, threw a carley float overboard, then slid or jumped into the sea. The skipper and the radio operator, Robert Donald Robertson of Hull, were still on the bridge. Robert Donald Robertson sent distress messages right to the bitter end. Both he and the skipper went down with the ship.

Eleven of the crew were on the carley float and two in the lifeboat. Unfortunately the next sea that came along overturned the lifeboat and dashed it against the side of the ship, throwing the two men into the water. The carley float was also flung over, scattering the men into the icy water. Only nine men remained on the float. They watched helplessly as one of their friends drifted away. David Atkinson, twenty years old and one of the youngest members of the crew, had been in the lifeboat. He managed to swim 100 yards to a floating log. There he watched as the *Hildina* rolled over and disappeared beneath the waves.

Some of the men including the mate, John Moran, and the bosun couldn't swim. They struggled to gain a hold on the lifeboat; the bosun held on with one hand and deckhand George Hayes clung to his ankle. The mate was first on the lifeboat and began to pull the rest on board. For the survivors the wait must have felt like a lifetime. The *Velia* was only 6 miles away when she picked up the distress call. Skipper Charles Pennington ordered his crew to chop away the gear. It took three-quarters of an hour to reach the scene; the upturned lifeboat was picked up by the ship's radar and helped them to be found.

When the *Velia* reached them the crew were already suffering the effects of exposure, some already unconscious and in need of urgent medical treatment. Some were too weak to grab the ropes being thrown to them. The skipper had to manoeuvre his ship close alongside the raft so each time it came up on a wave they could pluck the men off the raft and pull them onto the deck.

Deckhand James Benson was lost; as he was being pulled on board the ship he slipped through his lifejacket and tragically disappeared. David Atkinson was still clinging onto the log, wearing only his underwear and stiff with cramp. He secured his line with his teeth and was pulled to safety.

The survivors received medical attention by radio from a doctor back on shore. The ten survivors became nine as the cook died on his way back to shore. Several trawlers in the area searched for the missing crew and an aircraft was sent out from Kinloss. The *Velia* continued searching for several hours before leaving the scene. At 03.00 on the Wednesday they reached Stornoway.

We went to see if we could help with her mooring ropes or if there was anything we could do for the survivors, but they were well catered for by the Seaman's Mission and the Shipwrecked Mariners Society. The survivors were driven to the Mission for a well-earned rest. Around lunchtime we went to the Mission. I spoke to David Atkinson and the bosun, both full of praise for Charles Pennington, skipper of the *Velia*, though they didn't say a lot about what happened. None had serious injuries but they were all rather subdued and quiet, obviously still in shock.

Later that afternoon we, along with crews of several other trawlers that were in Stornoway, attended a service with the survivors on the *Velia*. The service was held by the Port Missionary, Mr Baxter. The coffin of Dennis Webb was lowered carefully onto the deck. After the service the *Velia*, having cut short her trip, left Stornoway with Dennis Webb's body on board and returned to Fleetwood. The survivors left Stonoway the following day. We left too and continued with our trip. I never met David Atkinson again after that. Forty years later I was skipper of a beam trawler *The North Sea*. On one trip a young man aged twenty-one joined us as mate – turned out it was David Atkinson's son.

I was away over Christmas on a trip. I wasn't keen on being away for Christmas but there were times you couldn't do anything about it. Our regular skipper had the trip off so the skipper who took us away was a young Ronnie Slapp. On Christmas Day we stopped fishing for a day off. I remember being in the wheelhouse, the skipper on the radio with other crews in the area, all taking it in turns to sing songs. The two deckhands, who both became top skippers out of Fleetwood, were asked if they'd sing a song; they refused, but the skipper talked them into it.

'OK,' said one of them, 'but you won't like it.'

'Doesn't matter,' said Ron.

They sang a song that they'd been making up all trip. Sung to the tune of *Answer Me* by Frankie Lane, it went something like this:

'Answer me Ronnie Slapp where did you lose that trap
Won't you tell us where you went astray please answer me please do
It was there yesterday
But when we hauled it had gone astray
Won't you tell us where you went astray?
Please answer us please do.'

I don't remember any more, though it was very good. The skipper, however, wasn't amused.

'You bastards are both sacked when we get in,' said Ronnie.

'Don't bother,' they replied. 'We're packing up any way.'

It was funny at the time but they all got over it.

The following trip I was told Arthur Lewis wanted to see me in his office. I was a bit worried about this – what could he want me for? I hadn't done anything wrong had I? When I arrived at the office he told me the skipper had said I'd done really well. Great, I thought, he was going to promote me to deckhand. I was surprised when he said he was taking me out of the *St Philip* and putting me to work in the net store! I didn't want to work ashore; all I wanted to do was to go to sea. He explained that mending nets a couple of months in the net store would do me good as I would see how trawls were made up in sections and how it all fitted together. So that was it, a shore job. The *St Phillip* sailed for her next trip without me; I was totally gutted.

The net store at Boston's was in the same building as the offices, in the basement but with windows at street level. I could look out and see the crews going to the cashier's office to get paid. I envied them but had to make the best of it.

The net store was a large room where a full trawl could be laid out. The two men working there if I remember correctly were two old skippers, Percy Bedford and Jack Chard. I think Jack had at least one son who was skipper in Boston's. Percy was from an old Fleetwood fishing family. I learnt about putting a net together and the different parts, such as top wings, lower wings, Belly and Baitings, cod ends and the square. Most important was how to put them all together to make a full trawl. There was a lot to learn but I was better for knowing it. It was the difference between having a heap of net on the deck in front of you and knowing in your mind's eye how it all fitted together.

One amusing incident that happened during my time there involved a big pit filled with tar. Our last job of the day was to take out the bundles of nets and hang them to dry over the pit, ready for use the following day. Then we had to fill the pit with new bundles of net to soak overnight. We were having a tea break when this deckhand staggered into the net store pissed out of his mind.

'Now then Percy,' he said. 'Just come to see how you're getting on.'

'Been in the pub again then Bill?' said Percy.

'Isn't it time you was going home?' said Jack.

'No I'm OK; going for another pint in a minute.'

With that he started to stagger a bit in the general direction of the tar pit.

'Keep away from that pit,' Percy hissed.

But Bill was a bit of a nosy sod and wanted to see what was in it. With that he tripped over some net and then proceeded to fall headlong into the pit.

'Bloody hell!' said Percy.

We dashed across to the pit and dragged him out. What a mess he was in; he was covered from head to foot in tar.

'Now what are we going to do with him?' said Jack.

'Best thing is to wipe his face and make sure he hasn't got any of that stuff in his eyes.'

We cleaned him up best we could.

'We can't send him home in that state,' said Jack. 'And I can't see any taxi taking him home in that state.'

Bill was still staggering about. 'F**k you lot, I'm off for a pint,' he said.

Jack took him out into the street and pointed him in the general direction of Dock Street and the pubs. Off he went, weaving his way down the street leaving a trail of black tarry footsteps behind him.

'I know his old woman,' said Percy, 'And she'll bloody kill him when he gets home.'

We laughed about it for days.

I spent three months in the net store and I have to admit I enjoyed it; I learnt a lot so it was worth it after all. However, I had itchy feet and was ready to get back to sea. Finally Arthur Lewis called me into his office, and asked how I'd been getting on. 'Fine,' I said.

He told me the two skippers in the net store spoke well of me and that I'd done OK.

'Are you ready for sea?' he asked me.

'You bet!' I replied.

'You'll be going back with Soccer in the *St Philip*,' he continued. 'But I want you to do a trip in the *Princess Royal* first.'

I was over the moon; the *Princess Royal* was a new modern diesel trawler. She was a great ship and had just come back from the Spithead Review. She had all mod cons, even a bathroom. She was only about two months old and all lovely and brand new. The living accommodation was roomy, clean and warm, there was no sea gear allowed anywhere near it. The bridge was the same, spotlessly clean and the brass work gleamed, in fact everything about her was spotless.

I'd hoped they'd let me stay on her, but no such luck, next trip I was back on the *St Philip* with Soccer. We did another couple of trips on her together; the firm was pleased with Soccer's earnings and promoted him to the *Mt Leonard*.

CHAPTER 7

ST LEONARD, 1954

I joined the *St Leonard* on 24 January 1954. A new diesel trawler just like the *Princess Royal*, she had all the same mod cons but was smaller. Diesel ships had a lot more power than old coal burners, caught a lot more fish and often did shorter trips but made a lot more money. The bosun was a chap called Peter Ince, a stocky man with a bark that was far worse than his bite. The chief engineer was Bob Dunks from Grimsby, a qualified diesel engineer in great demand as diesel trawlers were just making their mark on the fishing industry. He'd tell me all about Grimsby, the ships and the fish docks. He told me about all the notorious pubs down Freeman Street too.

The first trip went without a hitch; we landed an excellent haul of fish including 200 boxes of prime hake. The trip made just over £3,000, a big trip for 1954. With my poundage and backhanders off the crew I went home with £11 10s 6d. I'd never had so much money in my whole life!

When you're brassy it's your job to keep the forecastle clean and tidy, fill the coal lockers every day, make sure all the mending needles are full and well greased, and most important of all, keep the deckhands well supplied with tea and not to be cheeky. If you did your job properly, at the end of the trip the crew would slip you a couple of bob (a backhander). We'd go to Blackpool to celebrate; our favourite places where the Opera House to see a show or the Winter Gardens. We had a lot of good times in Blackpool.

It was my second trip in the *St Leonard* when something happened that scared the living daylights out of me. We were fishing; the weather was blowing about force 5 or 6, nothing that would normally bother us. This particular day we'd hauled our net and it had a couple of small holes in it. 'Mend it up,' the skipper instructed. The damage wasn't serious. It wasn't worth putting the port side net over. We lay to mend the starboard net. The skipper put the wind on our port side to give us a lee while we mended. As we were mending, the ship rolled to starboard and took a bit of water over the rail, which was nothing unusual, we just carried on mending. But she didn't roll back to port; she stayed with a starboard list. Suddenly she rolled to starboard again, this time filling the starboard side of the deck with water. We grabbed hold of the handrail underneath the bridge, now all up to our waists in water. We waited for the ship to roll back to port and to clear the water, which you would normally expect. She didn't, instead she lurched over to starboard yet again. Now it was deadly serious, we were laid right over and the sea filled the deck from stem to stern. Everybody made a dash to get off deck, which was difficult as the ship was now awash.

As we scrambled along the deck to get onto the casing one of the crew slipped, losing his hold on the handrail and slipping back into the water. He was being swept overboard but the bosun managed to grab hold of him and drag him back. The deck was awash from right aft all the way to the fore gallows.

The deck boards had all been unshipped and where floating about the foredeck. I was first onto the casing as I'd been stood alongside the ladder. The bosun followed me and luckily the rest of the crew managed to get safely onto the casing. The ship was over on her side at such an angle it was almost impossible to stand on the deck. We held onto anything we could to stop ourselves from slipping back onto the deck and into the sea. Peter Ince grabbed hold of me. 'If the worst happens you stay with me and I'll look after you.' I knew what he meant; I didn't need it spelling out.

A couple of deckhands were struggling trying to unlash a carley float; the ship was in such a position she was about to roll over. It was now a situation of hope and pray – another nudge to starboard and she would've gone right over and she would never have recovered from it. The minutes that followed seemed like hours, I looked round at the crew, all hanging on for dear life; they were silent, with anxious looks on their faces.

It was a great relief when the ship started to right herself, slowly at first, but she got herself back on an even keel. All the nets had been washed overboard and had to be pulled back in. There was something seriously wrong with this ship. I've sailed in many ships since but I've never seen one perform like that before. It certainly shook us all up. Several of the crew signed off that trip and vowed never to sail on her again. I never found out the reason for this behaviour as I left that trip to join my dad in Grimsby on the *Northern Duke*.

The *St Leonard* was sold to Canada and renamed *Zebra*; she was wrecked in December 1966.

CHAPTER 8

ST KILDA

From starting out on that first trip with my dad in the *Comitatus*, to leaving the *St Leonard*, I'd fished from Morecombe Bay, through the north and Channel, into the Irish Sea, along the west coast of Scotland, through the Minches and on to the Butt of Lewis. We'd worked to the north-eastwards along the 100-fathom line to the Shetland Islands as far as Muckle Flugga, the most northerly point of the British Isles. On the west coast of Scotland we fished round St Kilda and out as far as Rockall Bank which lies some 250 miles west of Scotland. This is a hellhole in winter, the weather can be stormy with hurricane-force winds, and being so far out in the Atlantic the seas build up to tremendous heights.

We went into some of the most picturesque harbours in Scotland, such as Stornoway, Oban and Tobermory in the Sounds of Mull, as well as Castle Bay on the island of Barra and Scrabster along the north coast of Scotland. In the Shetlands we visited the capital Lerwick and Scalloway on the west coast of the Shetlands. In Northern Ireland we visited Morville and Londonderry both in Lough Foyle.

We spent a great deal of time fishing around the island of St Kilda. To us it was just another island and not a very good place to work as the ground was very hard which meant a lot of net damage. One of my fondest memories of St Kilda was on a pleasure trip with my dad in the *Comitatus*. It was a glorious summer's day and the sea was like a millpond. He steamed the ship alongside Boreray and once at the nearest point he sounded the ship's whistle. What an unforgettable sight it was to see thousands of gannets take to the skies like a huge white cloud.

During winter the weather could be very stormy and consequently we spent a lot of time at anchor in Village Bay on the main island of Hirta. During my pleasure trips some of the older fishermen told me stories of when they'd take mail and stores out to the islanders. The older deckhands were born in the late nineteenth century and St Kilda was not evacuated till 1930.

I must admit I enjoyed my time fishing out of Fleetwood and often wondered if I should've stayed there.

CHAPTER 9

NORTHERN DUKE, GRIMSBY, 1954

After leaving the *St Leonard* I went to Grimsby with my dad. I was six-teen years old and had never been on a train before. Dad and I boarded at Fleetwood and changed at Manchester. Manchester Station was huge com-pared to Fleetwood and very busy.

I enjoyed the journey, it was brilliant. Steam trains are magic, an engineer-ing miracle and great to travel on. I've never had any inclination to be a trainspotter, but I can understand why they do it. In the '50s there were so many famous trains – the *Flying Scotsman*, the *Mallard* and many more – there were so many different types and each one had its own individual character.

Travelling by train that day, little did I realise I'd be travelling on them nearly every trip for the next five years! During the journey Dad offered me a cigarette. Surprised, I refused. 'I know you smoke Jim so you might as well have one,' he said. Then he asked me how long I'd been smoking. 'About six months,' I told him. I daren't tell him I'd been having a crafty smoke since I was about fourteen years old.

As we approached Grimsby Town Dad pointed out the dock tower, Grimsby's famous landmark. From that day on I always looked for it when retuning to Grimsby. Upon arriving we got a taxi to Boulevard Avenue where we had lodgings with Mr and Mrs Cavanagh. Mr Cavanagh was a retired trawler engineer and I lodged with them for the next year or so.

That evening I got the trolleybus (another first) to Cleethorpes and spent some time on the seafront. I walked back from Cleethorpes, down Isaacs Hill, along Grimsby Road to Park Street. As I crossed I was leaving Cleethorpes and entering Grimsby. The road changed name from Grimsby Road to Cleethorpes Road and I carried on to Riby Square. To my left I noticed the famous Freeman Street and the notorious Lincoln Arms pub that Bob Dunks had told me about. I crossed the railway lines into Victoria Station then on to the Palace Theatre, over Corporation Bridge into Corporation Road. Just at the beginning of Corporation Road was the terminus to catch a tram to Immingham. I carried on to Boulevard Avenue. It was a long walk. I couldn't believe the amount of pubs I passed; it seemed there was one on every street corner. Someone once said Grimsby and Cleethorpes were the last two places God made and that he had nothing left to put in them so he filled them with pubs and clubs. Maybe they were right!

Next morning a taxi took us Down Dock. We were due to sail about 11.00. The dock was a real eye opener, it was huge! It was like a town in its own right. It had its own bank, post office and labour exchange, numerous cafés and several paper shops. There were no end of ships' chandlers, net stores, riggers departments and shops supplying fishermen's clothing.

We were dropped off at the *Northern Duke* on the North Wall. I'd never seen so many trawlers lined up one after another, two or three deep in some places. Over 100 trawlers lined up from one end of the North Wall to the other. The deepwater trawlers were tied up near the lock gates, the Faroe and North Sea ships were tied up at the south end. There were more large modern trawlers than I'd ever seen, but at the same time there were a lot of old trawlers that looked about ready for the scrap yard.

The *Northern Duke* wasn't a modern trawler but an oil-fired steam trawler built in 1936 at Bremerhaven, Germany, as part of First World War reparations. Fifteen of these ships were built for Lever Brothers. They were: *Northern Pride*, *Northern Gem*, *Northern Wave*, *Northern Dawn*, *Northern Spray*, *Northern Foam*, *Northern Sun*, *Northern Sky*, *Northern Rover*, *Northern Chief*, *Northern Gift*, *Northern Reward*, *Northern Duke*, *Northern Princess*, and *Northern Isle*. The *Isle*, *Princess* and *Rover* were all lost in the Second World War. The *Chief*, *Gift* and *Rover* were sold to Iceland. The *Spray* was lost at Iceland and the rest were scrapped between 1964 and 1966. I sailed in the *Gem*, *Pride*, *Foam*, *Sun*, *Duke*, *Dawn* and *Sky*.

When I boarded the *Duke* everything was big, a bit different to the old *St Philip*. The agreement was I'd do three trips on the *Duke* with Dad and if I didn't like fishing deep water I'd go back to Fleetwood. I went ashore to get a few things to take to sea although I didn't really need much as the deepwater trawlers all carried the bond: this consisted of duty-free cigarettes and tobacco, soap, chocolates, matches and fag papers – it was like a mini shop. The ship also carried an abundant supply of beer, rum and whisky. There were several stores on the North Wall catering for fishermen and open every tide time day or night. You could get everything you needed from a new mattress to a new suit, and get it on tick. But they charged you 2s 6d and the prices were high. The money was stopped out of your pay at the end of the trip.

We let go of the ropes at 10.30 and made our way towards the lock gates. I was on the bridge as we passed through the lock gates; looking out the window the foredeck looked massive. We had to put a rope ashore at the corner of the lock gates to level the ship up with the lock; we let go the rope and made our way out into the River Humber. It was a clear day, the weather and visibility good. We navigated our way out into the river and altered course at Spurn Light Float to NNE. We were bound for the Norway coast.

At 07.00 the next morning all hands were called out for breakfast, then onto the deck to get a new trawl ready. At the end of each trip the trawl was chopped off and dumped over the side. The next trip would start with a new one, which I thought was a waste until I'd done a couple of trips. I realised these ships caught so much fish and the nets took such a hammering that by the end of the trip they were no good any way.

It'd take till twelve o'clock to get the work done. Any other jobs left were done by the day men in the afternoon. During the morning as the gear was being made ready the skipper dished out the bond. The rum was usually given out during time on the fishing grounds. Each man was given a dram of rum and a case of beer as he collected his bond. Some skippers dished beer out by the case whereas others only allowed you four cans a day. It really depended on the crew, and if they could be trusted not to drink it all at once. The majority of the crews were OK. But there was always one drunken bastard who'd mess it up for others. The biggest culprits were just a handful of skippers. Some were worse than most of the deckhands put together. I've sailed with skippers who remained permanently pissed until the booze ran out, and they got away with it. If the skipper was doing well and making good trips the owners turned a blind eye. But God help the crew if they got drunk. They could face being out of work for a month or two, known as the long walkabout. You'd be ashore until the runner decided you'd learnt your lesson. Only then would he give you a ship, but it would probably be one no one else wanted to sail in. She'd be a lousy ship with a skipper who didn't make any money. But you had to get back in the good books.

In the '50s, '60s and early '70s there were so many fishermen ashore at times it was hard to get a ship. The ship's husband, known as the runner, was in charge of manning the ships. If your face fitted or you dropped him a few bob each trip you were OK, but if you crossed him for any reason you had trouble finding work.

Life on the deepwater trawlers was hard. Once out the dock some skippers turned from gentlemen to right old bastards, driving the men and the ship to the extreme, working in all sorts of bad weather, never stopping to give the crew a break. The work was hard and the hours long. The conditions were atrocious and it was no good complaining to the office because they'd just give you the sack.

That first trip we worked the Norwegian coast. I wasn't on pay but Dad had me working as one of the crew and I did a full trip on deck for no money.

I found it hard going as I wasn't used to such long hours on deck. We had a break when we went into Tromsø (nicknamed the Paris of the north) for repairs. We were in for about thirty-six hours in which time a few of the crew got drunk and caused a panic ashore. They were returned to the ship by the police. We returned to the fishing ground and started fishing again. The fishing was good, but the weather had got worse.

At midnight I rolled in for my six hours' sleep only to be called out again at 03.00. Dad had told the mate to get me out to help out on deck. I wasn't very pleased but I daren't say anything about it. I'd done eighteen hours on deck had three hours' sleep and was facing another twenty-one hours on deck! Great, I'll be absolutely knackered by the time I see my bunk again, I thought; that'd be three hours' sleep in thirty-eight. This is the life!

I did one more trip with my dad to Iceland working at the East Horns, commonly known as the smash and grab. The ground there was hard as hell and the nets were torn apart every haul. It was the pits, very hard work and long hours.

On the way back to Fleetwood I told Dad I'd rather go in another ship as I didn't want the crew to think there was any favouritism. He agreed it was the best idea. When he left to go back to Grimsby I was a bit sad I wasn't going with him, after all I liked sailing with my dad and didn't mind the hard work, but you have to make the break and find your own way in the world at some stage. I guess this was my time.

Two weeks later I signed on as decky learner on the *Northern Chief*.

CHAPTER 10

NORTHERN CHIEF, GY 128, 1954

I joined the *Northern Chief* in the summer of 1954, a modern oil-fired trawler 692 gross tons and 178ft built in 1950 by Cochran's of Selby. Six of these ships were built for Northern Trawlers between 1949 and 1950. The skipper was Albert Meech. She was a good ship but a bit slow in bad weather. Head into the seas the speed was down to 5 or 6 knots, not a problem steaming to fishing grounds, but no good going home.

The deckhands still slept in the forecastle. Accommodation was good unlike the old *St Philip*. The forecastle in the bow was at deck level and divided up into two berths, one port and one starboard. Between these two berths was accommodation for twelve deckhands. The bathroom and toilet were on the starboard side. Another luxury! Unfortunately though when weather conditions were freezing the pipes to the toilets and bathroom froze, rendering them useless. The berths were fitted out with highly polished wooden panels and were reasonably comfortable, but like all ships when head to wind in bad weather, bloody uncomfortable!

The skipper's berth was under the bridge, done out to a high standard with wall lights and fitted carpets and its own private bathroom. The rest of the crew slept aft. At deck level was the galley and the crew's mess deck. The officers' mess deck was down below.

We had a good crew on the *Chief* and the skipper was one of the best I sailed with. As long as we did our jobs he left us alone. We worked mostly on the east coast of Iceland and concentrated on good quality fish from the whaleback area. The skipper wasn't a believer in chasing after fishing reports because when they came out the fish had usually been caught. The fishing was steady most of the time and what we lacked in bulk was made up for in quality. At the end of the day we made as much money as the other ships and had an easier life.

Steaming down to the fishing grounds we'd get the nets ready; we were usually done by dinnertime. The skipper would send a dram of rum down to christen the new trawl. The rum came on board in stone jars and had to be put into bottles. We carried plenty of rum; it was dished out a dram at midnight and one just before dinner. We also carried a couple of cases of whisky; this was for the officers to take home at the end of the trip.

After tea each evening we'd play cards in the mess deck. The skipper would play poker, nap or rummy. At the end of the session we'd have a mad half hour or so playing 'crash', a fast deadly game were you won a lot or lost a lot. There was no in-between. Those who didn't play cards would be in the forecastle having a few beers and a dram or two. The crew often took a bottle of rum or whisky to sea for steaming down. It was usually gone by the time we reached the fishing grounds.

One trip some crew made a homebrew by boiling sugar, yeast and potato peelings in an aluminium bucket. When ready it was put into lemonade bottles with screw tops and put in lockers to ferment. A couple of nights later we were woken up to the sound of bottles exploding. One of the crew jumped from his bunk and opened the locker. All the bottles had exploded except one. He managed to get the top off and the contents sprayed the locker door. As it ran down the door all the varnish came off with it. They decided not to drink it after all.

When we arrived off the Icelandic coast it was just after midnight, the weather was force 7 to 8 with a good sea running. We were called to get the net over the side. The skipper sent down our dram of rum then the watch below was set. This is the point when all friendship between the crew and skipper ends. In deep water trawlers the skippers ruled with a rod of iron. What he said went, whether we liked it or not. We didn't always like it but we did it. If he told us to work thirty-six hours we did it. Some skippers were evil bastards and if the fishing wasn't going well they'd take it out on the crew and find any excuse to keep us on deck. They'd have us making up nets and chopping away the ice. They generally f***ed us about when all we wanted to do was go aft, get our gear off and just rest for an hour or so. One of their favourite tricks was when the net was hauled up and was damaged they'd make us chop it off and put a new trawl alongside. The one we chopped off was dumped on the port side for us to mend in our spare time.

Our skipper wasn't like that and the crew was a good bunch. One of them, George, had been a rigger and was very good at splicing warps. He taught me how to do it; it wasn't easy but in the end I became very good at it. One morning we'd been practicing splicing and George was due to go below for his six hours. He told me to leave the splice alone till he was on deck again and we would finish it off. About 15.00 we'd got the fish gutted and down the fish room. I was a bit bored so decided I'd carry on with the splice. I put the spike into the warp to open up the strands and as I worked the spike down the wire the strands closed up, nipping my finger in between them. I tried to get my finger out on my own, but there was no way I could do it without help. I called one of the crew to give me a hand. He tried but couldn't do much about it so he went to get help. Now we had three of the crew trying to get my finger out but all they did was cause me more pain. As a last resort they called George out. He was still half asleep and in a foul mood. He called me all the stupid bastards under the sun, vowing never to show me anything ever again. He managed to free my finger but I am sure he caused me a bit more pain just to teach me a lesson.

A few trips later we'd just hauled our nets and let the fish drop on the deck. The cod end had been tied up and was hung outboard on the derrick. The weather was about force 7 and the ship was rolling a bit. The third hand shouted out to let go of the cod ends but the deckhand on the wire was a bit slow. By the time he let go the ship had rolled back to port. The hook swung in board, smashing him on the head. He dropped unconscious to the deck. The crew dragged him out of the fish and laid him on the fore hatch. I made my way foreword to see what was going on. I was sure he was dead. His head was covered in blood and he looked awful. I asked the mate if he was dead. Suddenly he jumped up and gave me a smack across my head, scaring me to death.

'No such luck,' he said. 'Now go and find my sou'wester. It's floating over the side somewhere.' F***ing old bastard, I thought, but I dashed to retrieve it anyway.

Albert the skipper liked his music and one of the joys of being in a modern ship was having music piped from the bridge to the deck. In Iceland the Americans have an air force base at Keplavík with their own accommodation and radio station. They sent out non-stop music twenty-four hours a day. The skipper tuned in and relayed it to the deck. It was a brilliant station, plenty of country and western music I'd not heard before and I soon got to like it.

The top two female artists at that time were Patsy Cline and Kitty Wells. Both were competing for the top spot in the American hit parade. The skipper had a radiogram in his berth which was fixed to the deck speakers. His favourite record was *The Wheel of Fortune* by Kay Star. If we'd had a big haul of fish, sure enough it'd be the first record played, and played more than once during over the next three hours till we hauled our nets up again.

Once when in the pound gutting fish with Kay Star belting out *The Wheel of Fortune* over the loudspeakers the ship suddenly lurched to starboard and the warps started pulling off the winch. We'd hit a wreck on the seabed. We rushed to get the gear hauled up, the skipper turned off the music and without warning a record flew from the bridge window and smashed to pieces against the mast. It was Kay Star. We'd only been towing our net for about an hour and a half. There was no damage and we had 100 baskets of good cod and haddock. After the fish was on board and the net back on the seabed the music came on again and guess what the first song was – Kay Star! I'm sure he had a locker full of that record!

When the weather turned bad, Albert would take us into the fjords. Two of his favourite places were Seydisfjord and Nordfjord. If the weather was really stormy we'd drop anchor.

Other times we'd go alongside and tie the ship up to the quay. The Icelanders were very good to us. The skipper was well known to them and had many friends. They'd open the swimming pool during the day for us and the dancehall during the evening. It made for a pleasant break during the trip. Often the skipper would have his music blasting out across the fjord. I'm not so sure how the locals liked it, especially late in the evening! Alas it all came to an end with the first Cod War.

We were home for Christmas that year and after the break I went down to the office to sign on. I was disappointed to find out I wasn't going back in the *Northern Chief* but highly delighted to be told I was going deckhand learner on the *Northern Jewel* instead. She was a brand new ship and the skipper was one of the highest earners. It was while in the *Jewel* that disaster occurred.

On 27 January 1955 two Hull trawlers, the *Lorella* and *Roderigo*, sank off the North Cape of Iceland. The previous day the trawlers had been fishing 90 miles north-east of the North Cape. The weather was very poor with gale-force winds, blinding snow and severe icing. There were other ships in the area, including British, French, German and Norwegian trawlers. The intense cold caused the ships to ice up. Two ships, the *Northern Crown* and *Northern Sea*, had to suspend fishing operations and spend hours chopping ice off their superstructure and decks to prevent the ships turning over. The *Northern Sea* returned to Grimsby. She was due to dock in the next couple of days.

The first indication of trouble came when a message received from the *Lorella* said she'd been thrown over onto her side and needed help. The *Roderigo* immediately went to her aid. In the early hours of Thursday morning she too sent out a distress signal. Nothing more was heard from either ship. An air sea rescue plane from the American air base at Keflavík flew over the area but could find no trace of the *Lorella* on the radar screen. Later the *Roderigo* also disappeared from the screen. The forecast for the area was north-east gales.

A huge search was mounted and ships and planes searched in vain for the two ships but nothing was found. There were no survivors from either. Forty men and boys were lost, leaving at least twenty children fatherless.

Skipper Tom Evens of Grimsby trawler *Stockam* heard other ships including the *Northern Crown*, *Northern Sea* and Hull trawler *Cape Spartel* talking about the serious amount of ice they'd accumulated. 'We had some icing up ourselves but we didn't have to stop fishing for it,' he said. 'I also heard three ships on the radio saying that weather on Bear Island grounds was as bad with severe icing and a gale of wind blowing, so they had to leave Bear Island and steam south to get out of it.'

With black frost the weather doesn't have to be blowing hard for ice to accumulate. It appears on the surface of the sea, like a low bank of fog. When it hits you the temperature plummets and ice starts to form on everything. When the weather's blowing and there's a lot of spray about the build up of ice is even faster. The only thing to do is get your gear on board and get out of it as fast as you can. If ice built up too rapidly we had to keep chopping it away to stop the ship from capsizing. Every fisherman dreads the stuff.

Another victim of the cruel conditions was the 656-ton Icelandic trawler *Egill Raudi* with a crew of thirty-four on board. She was stranded on the south side of Cape Ritur 20 miles west of the North Cape. The wireless officer said she was lying on her side, both lifeboats had been smashed and there was water in his cabin. The Grimsby trawler *Andaness* attended the scene with three Icelandic trawlers trying to drift life rafts to the stricken vessel. It was later reported the trawlers had managed to save thirteen of the crew and the other sixteen were taken off by the Icelandic shore rescue service. Unfortunately four Icelandic and one Faroese fisherman lost their lives.

CHAPTER II

NORTHERN JEWEL, GY I, 1955

The *Northern Jewel* was an oil-fired steam trawler and what a ship she was – fantastic! I can honestly say that I've never sailed in a better side trawler since. She was the best. The accommodation was first class and all the crew slept aft – far more comfortable than sleeping forward.

Our first three trips were to the west side of the Norwegian coast, at Anderness, nicknamed 'hands and knees', and that is what we were on at the end of the trip!

The skipper was Vic Meech, who was just getting started. The regular skipper was having three trips off. I liked Vic and I sailed with him in several ships but he was a hard taskmaster. Working the Norwegian coast we normally towed our gear for three tows one way. Then we had to get the nets on board and steam back to the place we had started. During the three tows we usually caught enough fish to keep us gutting on the deck until we got back to the starting place.

The weather was usually bad, so steaming back was mostly head to wind. It was a nightmare. Vic showed us no mercy; there was no such thing as doing a steady speed back, it always had to be full speed ahead into wind. The spray over the bow was heavy and every now and again the ship would dip her head into a heavy sea, filling the foredeck.

That's when the problems started. In decent weather the fish on the deck would stay put but when taking heavy spray and seas it wasn't so good. Water filled the fish pounds and everything changed. With the ship rolling from side to side, the water in the pounds meant the fish swilled about from one side to the other. The weight of the fish and water unshipped the pound boards leaving us with no boards in place. The whole starboard side was a heaving mass of fish, water and deck boards all washing about. We had great difficulty trying to keep on our feet as it was slippery with fish guts and livers and water nearly waist deep. The fish were washing over the after deck boards into the scupper and back into the sea.

Then Vic started. Down came the bridge window. He poked his head out and started screaming and shouting. He called us all the most useless bastards under the sun. Tempers were frayed and everyone was fed up. One of the crew lost it and started yelling back at him, telling him to get his f★★★ing head back in the bridge, but his voice was lost on the wind.

Vic eased the ship's speed down so we could get the boards back into place. Having got everything sorted out things got back to some sort of normality. Until Vic rang down to the engine room for full speed and it started all over again.

Skippers kept us up for long hours; twenty-four-hour spells weren't unusual. In the early '50s there was no such thing as a watch below. If the fishing was slack then we'd get some sleep but this came in short naps, anything from half an hour to three hours if we were lucky. The only time we'd get a decent nap was when we changed fishing grounds and did a steam for four or five hours. If the steam was any less than that we'd usually be repairing the net or sorting gear out. Only when the weather was bad and we had to stop fishing could we get a really good sleep.

The weather we fished in was horrific. We wouldn't stop for a gale force 8 and I've fished in a force 9 on many occasions. At times it took us all our time to stay on our feet. I don't believe we were aware of the true wind speed as we had no instruments to tell us. It wasn't until years later when I worked on an oil rig standby vessel with all the latest equipment on board that I realised what we must have fished in. I'd stand on the bridge watching the wind speeds reaching 40-50mph, and looking at the state of the seas, and I'd think, we'd still be fishing in this, we must have been crazy!

In the early '50s the ships started working a watch below. Some ships worked eighteen hours on and six off, which wasn't too bad as we managed five hours' sleep if we were lucky. Other ships worked twelve hours on and four off. This wasn't as good as you'd be lucky to get three hours' sleep out of that. Vic did very well on all three trips, landing around 3,000 10-stone kits of fish each trip. But boy we'd worked for it.

On the next trip the atmosphere on the ship totally changed. The regular skipper Bill Woods – Woody – was back. The crew was on their best behaviour – I think half of them were scared of him. When on the bridge taking our watch, we'd be chatting away to each other. Then the skipper would come in. All conversation stopped dead until he left. No one spoke unless he spoke to them first and if he asked for a pot of tea they just about fell over each other to go and get one for him.

I was never one for saying a lot but I didn't miss anything that was going on and it wasn't long before I started weighing people up. I soon found out whom not to tell anything to. If I wanted the skipper to know something I only had to tell certain crewmen and I could guarantee it'd get back to him.

During those first three trips the crew had told me about Woody, who was notorious for poaching inside the limits. They told me how when the ship was new, he had plates made to cover the name and number and painted the funnel so the gunboat couldn't recognise the company's colours.

At Iceland that summer most of the time was spent poaching and we caught a lot of fish. We worked two gears, one on the port side and one on the starboard side. The ground around the East and West Horns was very hard and our nets were ripped to shreds, but there was nearly always a good haul of fish, anything up to 100 10-stone boxes; on the other hand there could be no fish at all.

When we hauled and the net was damaged we always put the other net over the side. We had no time to hang about waiting for nets to be mended. The first haul the net was OK but after that we changed over nets every haul. Boy was it hard work. We never got off deck except for our watch below. It was eighteen hours of hard graft every day for twelve to thirteen days. By the end of the trip we were absolutely dead on our feet.

On the second haul the mate sorted the crew; some were to go gutting the fish and the others mending the net. We were on the starboard side as this was the side that needed mending when mate asked a deckhand if he could mend. I couldn't believe it. When I was in Fleetwood a deckhand wouldn't get a job if he couldn't mend. I told the mate I could mend but he just laughed and told me my job was to keep the needles full.

Once the crew was organised we set to work. I had to go with the menders and fill the needles. When all my needles were full I'd find a hole in the net and mend it up. I did this as often as I could to help get the job finished. The fishing started to get better and we had more fish than we could handle. The crew would go gutting while the radio operator Jack Douglas did the bridge work, towing the ship along. He'd been with this skipper a long time and the skipper trusted him.

The skipper himself was from Fleetwood and was excellent at mending nets very fast. I didn't go gutting but helped the skipper by filling the needles for him and mending any holes I'd find. The skipper noticed I could mend pretty well so started me on mending the bigger holes with him; in the end we did all the mending while the rest of the crew were gutting. I hardly did any gutting at all while at Iceland.

We used to work with Bill Drevers, skipper of the *Northella* from Hull. The *Northella* was on her maiden voyage that trip, brand new, the front of the bridge freshly grained. Our skipper was very impressed with her. When we sailed our next trip the bridge had been changed from white and had been grained in much the same way. Both skippers were from Fleetwood and good pals. Bill Drevers was as bad as our skipper when it came to poaching and they'd look out for each other.

One day we were fishing at Iceland and the weather was fine. We were with a group of ships fishing inside the limit and towing on a northerly course. Inside was a Hull trawler. I was on the port-side deck doing my usual job mending with the skipper, the rest of the crew gutting fish. Suddenly the bridge door flew open and the radio operator shouted, 'Skipper there's a ship steaming off the land and heading at full steam towards us'. No one had realised the gunboat had sneaked close in along the land and was now 3 miles from us and the other ships. Someone was going to get caught.

All hell broke loose. The crew dashed out of the fish pounds and got to their stations for hauling the net, the winch heaving at full speed. The crew had done this so often they had it down to a fine art. When the net appeared on the surface it looked like we had more than one bag of fish. The net was franticly scrambled in until we had enough net inboard to get a rope around it. A couple of heaves on the rope were enough for us to grab the wire leg that split the bag of fish into two sections. No sooner had we got the Gilson wire into the becket than the skipper rang the engine room for full speed.

Up until now it'd all been straightforward. From here we had to be very careful. We had one bag of fish hung up in the pound; the middle section of the net was still over the side and the ship was now steaming at full speed with the skipper screaming out of the window to get the rest of the net aboard. The net was streaming aft, the trawl doors were crashing about, and it was dangerous trying to get the gear aboard. Fortunately the net over the side split and we lost the fish, making it easy to get in. The most dangerous part was trying to hook the Gilson into the after door as it was swinging and crashing about; we had to be careful not to get trapped between the gallows and the door: 1.5 tons of wood and steel can cause serious injury or even death.

We eventually got all the gear on board safely. The gunboat was now right in among the ships. We heard his gun go off and I ran round the other side to see what was going on. The gunboat was alongside of the Hull ship which had not seen him soon enough. Our skipper was certain we'd been the gunboat's target but it had decided to go after the nearest ship instead.

We steamed for a couple of hours, listening to reports telling us no other gunboats were in the area. The Hull ship had been arrested and was being escorted to Reykjavik. Our skipper simply turned the ship around and steamed straight back into the limits! There's no accounting for cheek.

Skipper said we'd done a good job and sent an extra dram of rum down for us. We only stayed inside the limit line for the time it took the gunboat to get to Reykjavik and back. When it arrived back on the scene all ships were towing their nets well outside the limits as if nothing had happened. The gunboat had got his man and we'd had a bloody good day's fishing while he was gone so everyone was happy, apart from the Hull skipper who'd been caught. We spent most of our time within the limit. It was said if it hadn't been for fishing inside the limits our skipper wouldn't have been as successful, but then fortune favours the brave they say.

On the next trip the weather was beautiful, the sun was shining and you could see for miles. We were steaming into the land and the crew was on the deck getting the gear ready for fishing. I was on the bridge steering the ship while the skipper and radio operator were looking through the binoculars at a ship that was inside the limits. He was just laid to and not moving. They were sure it was the gunboat. 'We won't be going inside today,' the skipper said, to no one in particular.

As we got nearer the limit line it became clear it wasn't the gunboat at all but the Grimsby trawler *Vanessa* laid just inside the limit line. We steamed up to see what was going on. As we got alongside we could see her decks were full of fish. The two skippers had a few words about the fishing and we steamed another ten minutes, easing the speed down. The skipper was watching the seabed on the echo sounder. The sounder showed huge amounts of fish but the ground looked very hard and full of sharp pinnacles. 'I'm not too sure about that ground,' he said. 'But we'll give it a go.'

After we'd got the net on the seabed we went aft for a pot of tea and a fag. We were sat in the mess deck having a yarn when the ship's telegraph rang signalling we'd be hauling in ten minutes. This was unusual as we'd only been towing the net for an hour and three quarters, instead of the usual three hours. We knocked the warps out of the towing block and started to heave up. Everyone was waiting for the trawl doors to come to the surface but there was still about 75 fathoms of warp to hove up. Suddenly the water started to bubble up and turn a deep blue. With that the cod end burst to the surface and shot out of the sea like a giant haystack; it rolled over and spread out on the surface of the water like a huge sausage full to the headline with cod.

'Ease the winch down!' yelled the skipper. 'Heave away slowly!'

I'd never seen so much cod in one haul; we could see them swimming out of the mouth of the net. It must have taken two hours or more to get the fish aboard. The result was thirteen bags of fish on the deck, each bag containing approximately thirty 10-stone boxes of fish, a total of about 390 10-stone kits of fish. In a word, amazing! The starboard side of the decks was full from the winch to the fore gallows.

We shot the trawl again and started gutting the fish; everyone was in a good mood and chatting away to each other. Due to the amount of fish we'd caught on the last tow the skipper decided to cut down the time, and only towed the net for one hour and ten minutes.

As we hauled up a second time everyone watched the sea for the telltale sign we'd scored another big haul, and sure enough the net burst to the surface once more, straining with the weight of the fish. This time there wasn't quite as much as before, only eleven bags of cod equal to 330 10-stone boxes, though still a huge haul by any standards. We had to get more deck boards shipped up to stop the fish from going into the scupper. The starboard side and the middle of the deck were now full. I'd expected we'd now lie to and get the fish gutted but the skipper had other ideas: we shot away again.

Bloody hell, I thought, if we got another big haul where's it all going to go? We'll never get it on deck. After we'd shot the gear away the skipper ordered everyone to start throwing fish onto the port side as fast as possible. This was hard work as they were all big and heavy, some weighing up to 3 stone. The sun was shining and it was a warm day, it was energy sapping! We filled the port side of the deck level with the ship's rail. We then filled the middle of the deck and forward.

The starboard side was now nearly empty – but not for long. The skipper ordered the nets up straightaway. The net had been on the seabed for three quarters of an hour and produced another nine bags of fish, equivalent of 270 10-stone kits of fish. We only managed to get six bags on deck and even then some spilled over the rail back into the sea. The seventh bag brought on board wasn't released but left hanging on the end of the Gilson, the rest left over the side in the net. We dropped the scupper doors and had to clear the fish from around the fish room hatch. The only place we could put the fish was on the after deck.

It was estimated that in ten hours we'd caught 10,000 stone of fish. The net was only on the seabed a total of three hours forty minutes. We spent the next twenty-four to thirty hours laid to gutting fish. When we'd about thirty kits of fish left the skipper instructed us to shoot the trawl away. That thirty kits left was enough to keep us gutting until we hauled again. No rest for the wicked!

We kept this fishing going for the next three days. The weather stayed fine, which was a big help. When we left Iceland our decks were absolutely full. The skipper stopped the watch below, it was all hands gutting the fish – twelve men gutting non-stop for twenty hours. I was lucky as I'd just had my watch below when the skipper stopped the watches. The three men who should have gone below after me finished up doing over forty hours without sleep. The poor bastards were totally knackered, they didn't know what time of day it was. But it'd been a hard trip for us all.

The last fish went in the washer as we were abeam of the south end of the Faroe Islands. Out of the four days fishing we only had about twelve hauls, the rest of the time we were laid gutting. We finished up with over 3,500 10-stone boxes for four days' fishing, every one caught inside the limit line. The roundtrip was ten days, one of the shortest trips to Iceland by a Grimsby trawler at that time.

We were expecting a big grossing when we landed, but unfortunately several other trawlers landed big trips that day. The quality of our fish had suffered from being on deck for long hours in the hot sun and some of it was condemned. The rest made very poor prices; a lot of hard work for nothing. We were bitterly disappointed.

Our skipper was always in the limits but was very lucky. I remember one particularly dark night there were a few ships working together in the limits and no one saw the gunboat until the last minute: he was only about 3 miles off when he was spotted on the radar. Panic ensued. The usual routine was when we'd got the gear on board all lights were switched off, including the navigation lights. The ship in total darkness, we steamed straight out from the land. As we left we heard the gunboat firing his gun. He'd caught someone.

We'd been towing our net for three hours and were due to haul when the gunboat appeared on the scene. All we had in the net for our three hours was three or four baskets of fish. We kept steaming for about two hours until the skipper was sure we weren't being followed then we shot away our net. Three hours later when we hauled we had 100 10-stone boxes of fish. If that's not luck I don't know what is!

Fish weren't always plentiful and there were times when it was very difficult to find. This particular trip was one of them; no matter where we went, inside the limits or outside, there wasn't much fish to be had. Ships were going home with very poor catches and reports from all the ships were the same. We'd three days' fishing left and we only had 750 kits of fish. Normally at this stage of the trip we'd be looking at 1,700 to 2,000 kits in the fish room. Woody decided to steam to the Faroe Islands with the intention of having twenty-four hours' fishing. If there was no fish we'd steam to Grimsby and cut our losses.

We arrived at the Faroe Islands in the early hours of the following morning and shot our gear away on Fuglo Bank. We weren't very hopeful and we all thought that it was a waste of time, but Woody's luck was still with him. On the first haul we had 200 baskets of cod and coleys and three days later we were on our way home with 2,500 kits on board. We made an excellent trip due to the fact other ships had had poor trips.

The following trip we were back at Iceland on a trip without incident. Everything ran smoothly and fishing was good. We caught about 2,500 kits. Then the fishing slacked off. We only had about twelve hours' fishing time left so Woody decided it'd be a good idea to steam to the 100 fathom line off the north coast of Scotland in search of dogfish. Everyone hates it when skippers do this. It's nice to finish the trip at Iceland and steam straight home from there. It wasn't as if we'd a couple of days' fishing left. We only had twelve hours. We arrived on the 100 fathom line and shot away the trawl, just enough time to get three short tows in.

We called Woody all the greedy bastards under the sun; we were hoping there was no fish as we were ready to go home. We hauled after three hours. The weather was good, the sea flat and calm and we watched the sea for the telltale sign of bubbles that indicated we'd a good haul. Not a bubble in sight. Looks like there isn't any fish, I thought to myself.

We heaved the net to the ship's side. As the first part appeared out of the water I could see there were one or two dogfish hanging in the trawl. My mind went back to the old *St Philip*, catching a load of dogfish at Barra Head. My heart sank. It was now clear we'd a large haul of dogs. Dogfish differ from fish like cod and haddock as they've no swim bladders; there's no buoyancy so they don't float. It's like having a net full of rocks. It was hard work trying to get them in: every time we had so much of the net inboard we had to get some strong chains and lash it to the ship's rail, and even then the chains were in danger of parting. Had there been any swell the sheer weight of the dogfish would've ripped the net off. We had to get a rope and try to get it threaded between the rail and the net. We then made a loop and dropped it down the net, the rope then taken to the winch and heaved on. This was done very slowly, the rope could easily have parted, but we managed to get it up far enough to grab the line that's attached to a becket. This allowed us to lift the bottom section of the cod ends on to the deck. When we had the first bag of fish on deck the job should have been so much easier. But it wasn't! As we were heaving the next bag the becket round the cod end snapped; the whole net fell back into the sea and sank like a stone. Now we had a problem. We tried without success to get a rope down to the bottom part of the net but failed every time.

The skipper decided to heave the fish aboard in one go. We heaved the net up as far as we could, then put a strong becket round it. We got the tackle, a block with a three-purchase wire and a large hook on it, similar to that you see hanging on the end of a crane; it's very strong and takes two or three men to drag it along the deck. Having got this hook into the becket the skipper ordered everyone off the foredeck in case anything parted. We proceeded to heave on the tackle. I was stood on the starboard side out of the way and watched as the net was very carefully heaved out the water. As the tackle took the full weight of the fish the ship heeled over to starboard and the net came slowly out of the water. The block had now reached the top of the mast but the end of the net was still over the side of the ship in the water.

Suddenly without warning the tackle wire parted. The whole lot crashed back into the sea and the block hit the deck with a mighty crash, smashing one of the deck boards in half. Thank goodness there was no one stood under it. We rove a new wire up to the tackle block on three separate occasions. Each time the wires parted. We'd now spent six hours trying to get the fish in and everyone was well and truly fed up. We'd run out of tackle wires but the skipper wanted one more go; if it didn't work then we'd chop the net away and let the whole lot go.

The next order was to get a messenger wire up from the forehold; this is a single wire but thick which we rove on to a single block at the top of the mast. I thought he was crazy! Having parted four tackle wires I couldn't imagine a single wire lifting all that weight on board. We started again and heaved the net up with the rope, put the becket round the net and the hook into the becket. Again we started to heave on it. Everyone got out of the way and I went back to the starboard side to watch it being hove up once more.

As the messenger took the weight you could hear the grinding and groaning from the top of the mast; any minute now it was going to part the wire or pull the block from the top of the mast and I feared the worst. The net came slowly up the side of the ship like a huge fish sausage. Bit by bit it inched its way up till the messenger hook was tight to the block at the top of the mast. The winch was stopped but there was still about 6ft of it over the rail.

The skipper ordered the mate to get a rope and make it fast to the cod end knot at the lower end of the net. This was difficult and dangerous and the mate had to lean over the rail to do it. The problem was he couldn't reach it. The block at the top of the mast was still creaking and groaning – if the wire parted he'd go over the side with the net. The mate called two deckhands to hold onto his legs. When they had a good grip on him he wriggled himself over the rail until he was hanging upside down over it, the two deckhands hanging onto his legs. It was a good job he was a decent mate, or they might've just let him go! After a struggle he managed to get the rope fast and it was taken to the winch. The winch was put on its slowest speed until the cod end was pulled aft far enough to roll the end of the net inboard. As it hit the deck the block at the top of the mast came crashing down – the roller inside the block had shattered into hundreds of pieces. He was one bloody lucky mate!

In that last bag of fish there was about 160 10-stone boxes of dogfish weighing about 10 tons. When we took out the first bag there was about thirty 10-stone boxes in it, and we put 190 boxes of dogfish on the market; that's a lot of fish to heave on board on a single block and wire.

In June that year I had to go for a medical examination for the Royal Naval Reserve. It was discovered I'd a double hernia. I knew this as I'd had it for many years and had often told the doctor about it but they never found anything. After the examination I went back to sea. We were halfway through the trip when the skipper called me to the bridge and told me my dad had informed the office of my medical condition. I was put on light duties for the rest of that trip. On returning home I went into hospital for an operation. After my recovery I went back on the *Northern Jewel*.

We had a new deckhand with us that trip. Apparently he'd been the deck-hand learner while I was in hospital. I was a bit put out about it as I thought the skipper would've given me the deckhand job first. I didn't think this was very fair. I stayed with the *Northern Jewel* till Christmas then asked Tommy Rowson, the ship's runner, if he'd give me a start as a deckhand. He agreed so I signed off the *Northern Jewel* to go deckhand on the *Northern Gem*, not a good move but you have to start somewhere. I enjoyed my time on the *Northern Jewel* and although I got on very well with Bill Woods our paths crossed a couple of times during my career and it wasn't for the better.

CHAPTER 12

PRINCE CHARLES, 1955

It was December when I left the *Northern Jewel* so I decided I'd have Christmas and New Year at home before going as deckhand. On 24 December 1955 the *Grimsby Evening Telegraph* reported the sad news that the Grimsby trawler *Prince Charles* had run aground on rocks off the Norwegian coast. She'd sailed from Grimsby on 3 December and ran onto rocks off the island of Soeroeya, 60 miles west of Hammerest, Norway, during a snowstorm in the night. There were two dead, seven missing and eleven rescued. One of two Norwegian pilots aboard was missing but the other one was safe. The tragedy meant a grim Christmas for many families. Of the dead and missing six were from the Grimsby area and three from Hull.

The survivors had clambered onto a rocky islet and were rescued two hours later by the Norwegian ship *Ingoe* which took them to Hammerfest. The *Ingoe* also landed the bodies of the two dead. Seven of those saved where treated in hospital for frostbite and exposure while the others were taken to a hotel.

Prince Charles was owned by the St Andrews Steam Fishing Company Ltd of Hull and managed by the Boston Deep-sea Fishing Company of Grimsby. A spokesman for the company said *Prince Charles* had a good catch on board and was returning to the Humber from the Barents Sea. 'It was a cruel blow to have finished work which was difficult and dangerous enough and then to be lost within sight of the land with two Norwegian pilots on board,' he said.

Reports said the radio operator Osborne, chief engineer Pomows, cook Smith and galley boy Howard had been taken to the Grand Hotel in Hammerfest. Mate Hopper and deckhands Kershaw, Proctor, Haynes and Stevens were reported to be in hospital suffering from mild frostbite. Hopper also had an injured foot. Decky learner Sumpton and deckhand McGill were believed to have pneumonia.

In a statement the Grimsby steam fishing vessel's Mutual Insurance and Protecting Company Ltd said its agents in Hammerfest reported no hope of finding the seven missing men alive.

The *Prince Charles* stood on her bow totally submerged with 2 fathoms of water over her stern. The wreck was marked by a lifebuoy 160 yards east of Karken light in Soroey Sund.

The Norwegian naval frigate *Nordkyn* remained at the scene all night in the hope of finding survivors. She was joined by the Hull trawler *Kingston Topaz* but the search was called off at 07.00. During the night the searchers reported two empty lifeboats and a trunk had been picked up. Searchers said if any survivors had managed to escape they'd have come ashore. It was feared the wind and tide may have carried them out to sea.

Prince Charles was one of the most modern trawlers. She had been completed three years earlier and was fitted with all the latest safety devices. A Norwegian salvage vessel was on its way to investigate the possibility of salvaging her.

My dad was on his way home from fishing at the White Sea in the *Northern Duke* when the *Prince Charles* ran ashore. He was instructed to proceed to Hammerfest, pick up the bodies of the two crewmen and proceed to Grimsby. It was their final voyage home.

On board the *Northern Duke* was a deckhand named Jimmy Trew on his last trip to sea. On his return he was retiring at the age of fifty and leaving the fishing industry to work ashore.

When I'd first come to Grimsby and sailed in the *Northern Duke* with my dad, Jimmy Trew was one of the deckhands. He used to keep a diary of each trip. This is what he wrote about his last trip in the *Northern Duke*, a fitting tribute to the men who'd lost their lives on the *Prince Charles*.

The Last Voyage
By Jimmy Trew

In the days of my youth back in the 1920s, I first began my sea career as a deep-sea fishermen sailing from the port of Grimsby in a North Sea trawler, which was one of the fleet owned by Sir Thomas Robinson. Sir Thomas's father had been a successful skipper who had invested his money wisely to buy his first trawler. From this small beginning was built a large and well-founded fleet of trawlers which fished the North Sea grounds very successfully, and some of them in later years began to venture further afield to the Faroe Islands and the Shetlands. In those days fishing and its various subsidiaries provided jobs in Grimsby which was becoming the biggest fishing port in of the world.

Like many of my contemporaries I went to sea shortly after leaving school, first of all going for what was laughingly called a 'pleasure trip' were I was on the receiving end of all the usual horseplay, and where I stubbornly refused to take any notice of the hard-bitten old seadogs who told me that rather than go to sea for pleasure, I should go to hell for a pastime.

For many years I was content to remain as a North Sea fisherman or meggying as it was called, and indeed there is no finer breeding ground to learn the art of net mending and all other various skills which go to make up a trawler man. But eventually with my imagination fired by the tales I heard from my shipmates when we got a yarn bent on, about the faraway places with strange-sounding names were they had fished, I got the wanderlust and began to sign on the much larger deep-sea trawlers which fished in the tall waters of the Arctic.

Over a long period which lasted until I swallowed the anchor and came ashore at the age of fifty, we fished in such outlandish places as the North Cape of Iceland and the infamous Denmark Straits were so many ships have iced up and sank with all hands, Bear Island and further north at Spitsbergen – when it was not ice-bound – and the White Sea as far east along the Russian coast as Novya Zemlya, and even in the stormy waters of Cape Farewell at Greenland. This is a particularly inhospitable region, for one encounters the huge icebergs which are swept down to the Cape by the Greenland currents and into the Davis Straits. It was on one of these trips that the happenings of which I am writing took place.

We sailed from Grimsby in the Northern Duke on 10 December. Our spirits were rather low as we knew that we would be spending Christmas at sea – a daunting prospect but we had no time to brood as we were all kept busy preparing for the trip ahead as we gradually made our way to the north north-east and into the Arctic Circle. The skipper had decided to take us through the fjords which were usual in the winter time, and even before we entered the fjords we began to ice up as we encountered the dreaded black frost, as it was called. We were not in any danger of course, so we were not alarmed, but nevertheless it seemed to be a bad omen, as none of us had seen the ice as far south as this before.

We pressed on, picking up the pilots at Lodigen, and when we stopped at the other end of the fjords some thirty hours later at Honningsvag to drop the pilots we were in a sorry state – iced up solid from stem to stern in the perpetual Arctic darkness. There was no daylight now and prospects were gloomy. But we made our own way out of the fjords passing the North Cape, and then we stopped and lay under the lee of land and chopped away all the tons of ice with axes, heavy crocodile spanners and hot water hoses etc. until after several hours of hard work we were once more seaworthy. Then we were able to steam off clear of the land where we found we were also clear of the black frost and we began fishing operations. I will not dwell too much upon the many misfortunes which fell upon us, suffice to say that nothing seemed to be in our favour. There was hardly any fish to be caught and almost every time we hauled our net was torn and split, so we were continually at work mending the nets. The weather also had worsened and we began to encounter gale-force winds from the north-east.

On Christmas morning at 02.00 we had to scramble our gear on board and lash everything up securely and commence to dodge, head to wind in a force 9 gale, accompanied by fierce snow squalls which cut the visibility down to a few yards. Because it was supposed to be the festive season we thought of all the lucky people ashore enjoying themselves and began to feel sorry for ourselves, thinking that fate was against us as the gale continued for another day. Then we heard over the radio that one of our sister ships the *Prince Charles* had been lost in the fjords with some loss of life. It made us realise how lucky we really were – at least we were still alive and in a well found ship even if conditions were a bit grim. And our sorrow then was for our lost comrades, whoever they were: eleven had died, seven were missing, two were saved and some of us were bound to know them as shipmates.

Eventually the storm abated and we continued our fishing operations, but still without much success, unfortunately. By the time we started on our homeward journey we had only caught about 1,400 kits of fish, which was not very good. But other ships also reported small catches so we were not alone. We put into Honningsvag to pick up a pilot and there we received a message from the owners telling us to proceed to Hammerfest to pick up the bodies of some of our comrades for the passage home, and this, of course, we did with the utmost dispatch. We made fast to the quay at this far northerly port at about 10p.m. on New Year's Eve and I shall never forget the scene if I lived to be a hundred.

It was a dry, bitterly cold night, with a big bright moon shining above, and the twinkling lights of hundreds of stars shining in the sky. Everywhere was covered in ice, as it always is here at this time of the year, but it all looked so beautiful as it glistened and shone in the bright moonlight, hard – like glass – and almost crackling in the dry bitter cold of this lovely Arctic night. And to complete the picture, and indeed the most glorious and unforgettable part of the whole scene, was the magnificent sight of the Northern Lights in all its glory sweeping back and forth across the heavens like huge beautifully coloured searchlights – a sight to be treasured for always. I have seen the Northern Lights many times, but never like this, and I will never forget them.

But we were here on an errand of mercy, and so a party of us went ashore and we were directed to a small wooden church where we were introduced to an elderly Reverend Gentleman who was Norwegian but spoke good English. And here we saw the two bodies of our comrades in two plain white coffins in which they were to make their last voyage home to their grief-stricken loved ones. And we prayed with these strangers from another land for our dead friends and sang a hymn – and mine were not the only eyes which were damp. It is this time and place I remember when I hear the lovely words of *Abide with Me.*

We lifted our comrades' coffins high on our shoulders and bore them down to the quay and aboard the ship. Every Norwegian we passed on that bitterly cold night bared his head respectfully in a last salute. We set forth once again for home, and again there was a feeling of gloom on the ship as there always is in the presence of death. Our skipper Bill Greene who had served through the war like many more of us had seen death many times and gave all hands a pep-talk, and spliced the main brace, telling us to snap out of it.

Our troubles were not over on this ill-fated voyage for after we had cleared the fjords and were steaming southerly along the coast the wind freshened up from the north north-east and very soon we were running before a force 9 gale with mountainous seas piling up astern, threatening to swamp us. It really was a fearsome sight when my watch mate and I took over the watch for the afternoon with the wind howling through the rigging and the ship veering violently in the huge following seas as we fought to avoid broaching to it.

Then the inevitable happened: a tremendous sea caught us as it roared up like an express train and broke aboard us engulfing us entirely from stem to stern. It was an awful moment as we strove to keep the ship steady, and when a second sea smashed aboard I for one thought we were going to run under. For as we looked ahead all that could be seen was the foremast until very gradually we felt the ship begin to shudder and tremble beneath us, and slowly her bows emerged steaming from what could have been our watery grave. We could breathe again. Our skipper then deemed it prudent to turn into the storm and dodge, which we did for the next twelve hours.

The following day when some of the crew was at work in the hold fixing trawls, one of them, who was sitting by one of the coffins filling needles, looked at it and said, 'There is one thing old chap, you're certainly giving us a stormy ride on your last voyage,' which proved that the tension was now over and things were back to normal. For life has to go on; in trawlers all the trials and troubles of the trip are forgotten once you are homeward bound. The rest of the homeward journey passed uneventfully. Before we docked the two coffins was brought up onto the deck and draped in the Red Ensign under which they had sailed. As we passed through the Loch Pits everyone bowed their heads in salute to our flag which was at half mast.

I went Down Dock and stood at the lock gates. A huge crowd stood silently with their heads bowed as the *Northern Duke* with the two coffins on her foredeck passed slowly into the dock. Dad gave me a wave from the bridge, which I acknowledged. I didn't go round to the fish dock to see the coffins brought ashore – it would be an emotional time for the families and I didn't want to get in the way.

CHAPTER 13

STARTING AS A DECKHAND, 1955

I was embarking on my first trip as deckhand on the *Northern Gem*. I got on pretty well in her. The skipper's nickname was Bedroom Tommy as he spent the majority of his time rolled in. I only did the one trip as I was relieving one of the regular crew.

My next ship was the *Northern Dawn*, a coal-fired steam trawler. When we left dock the foredeck and half the fish room was full of coal to be shovelled down the fish room after we'd made enough room to accommodate it. As we sailed out of the river and turned our course to the north north-east we found ourselves head to wind in a storm force 10. After two hours steaming the coal on deck had been washed over the side by the huge seas pounding the deck. I can't say I was sorry to see it go!

The skipper was from Hull and I didn't like him very much. In my opinion he was a bit of an a★★★hole. By the time I'd done three trips I knew I was right. He didn't seem to have a clue. All he talked about was his garden. Maybe he should've stayed at home and looked after it.

I didn't like the ship either as there was no fish washer so fish had to be washed by hand. We'd spend twelve hours gutting the fish and throwing them across the deck onto the port side, hard work as some weighed as much as 2 stone or more. When the port side was full they had to be washed individually by hand. The donkey (water hose) was put into the fish pound until it was awash, then we'd jump in and start kicking the fish about, pick each one up, thoroughly wash it and throw it down the fish room. It took ages and by the time we'd finished we were absolutely soaked through and freezing cold. At times we must have been very close to hypothermia.

We were working on the East Bank and catching lots of duff. They call it duff because it's like a sponge or suet pudding but very heavy. When you broke it open it'd be filled with lots of little spikes like shards of fibreglass; very sharp they stuck into your skin like needles and chafed all the fish.

Each haul we were catching 100–120 10-stone boxes of fish with heaps of duff amongst it. In most ships you'd sort the fish out and dump the duff back over the side, so you'd only the fish to start gutting. Our skipper hadn't the patience to spend time sorting. We dumped all the fish and duff on the deck together. With the weather bad and the ship rolling about the duff chaffed and crushed the fish.

We finished up with a good trip but when we landed the fish it was in a very poor condition. Consequently the biggest part of the trip was condemned or left unsold. The trip didn't make enough money to cover expenses so we all landed in debt and had to get a sub off our next trip's money. I signed off and didn't go back. After all nobody wants to work for nothing.

Three-day millionaires

Landing day was always very busy. From midnight the lumpers started to land the fish ready for the sales which started at 07.30. At 06.00 the fish merchants were down, inspecting the fish and deciding which they hoped to buy. By 08.45 all the fish would have been sold.

We'd start arriving at the office around 10.00 to collect our settlings: having been at sea for three weeks we'd have three weeks' money to collect. It varied from nothing on a poor trip to a substantial amount on a good trip. Deckhands' pay was about £6.50 per thousand, so on a £10,000 trip his poundage would be £65 plus cod liver oil money. Cod livers were saved throughout each trip to be processed into cod liver oil.

It was an awful lot of money back then but it was usually gone by the time we sailed again! From the office the first stop was The Humber pub. A few pints there but we didn't stop long, too many scroungers. If there were two or three of us we'd order a taxi to run us around and he'd stay with us all day. Next stop was Freeman Street. It was party time all day and every day in Freemo; there were over 200 trawlers in Grimsby at that time landing fish six days a week. The Lincoln Arms was a regular haunt of the girls of the night (and day). They knew which ships made the best trips and who were the best spenders. Across the road was Cotties Bar, very similar to The Lincoln. Everyone here was well dressed. We'd all wear suits and ties; none of the off the peg stuff, it was all made to measure back then!

Around 5.00p.m. we'd go home and get something to eat. The taxi was paid with instructions to pick us up again at 8.00p.m. and we'd head back down Freemo for a few pints and then on to one of the dancehalls. The best was the Gaiety but there was the Pier at Cleethorpes and the Café Dansent, a great place especially when they had the Blue Bus dances. There was another place called Costello's above a shop in Victoria Street. My brother Bill and me decided to go one night. There was a door on the street leading up to the dance floor. As we arrived there was a commotion at the top of the stairs. Two bouncers had hold of a girl and they threw her from the top to the bottom of the stairs. She was well p★★★ed but being gentlemen we helped her anyway.

CHAPTER 14

NORTHERN SKY, 1956

I joined the *Northern Sky* the following day. The skipper was Vic Meech, the same skipper as on the *Northern Jewel*. It was his first regular skipper's job. He was a hard taskmaster and worked his crew hard, understandable as he was trying to make a name for himself. The mate was a friend and neighbour of mine, Johnny Meadows, who would later became one of Grimsby's most successful skippers, finishing up in Grimsby's top trawler the *Ross Revenge*. During his time as skipper Johnny broke the port's earnings record several times.

We were working at Iceland and getting a lot of nets ripped apart owing to the rough nature of the ground. We were on deck eighteen hours most days and it was pure graft for most of the trip. A lot of deckhands didn't like it and signed off at the end of the trip. It was the mate's job at the end of a trip to inform any deckhand who'd got the sack. He didn't sack many as most of them signed off.

I liked sailing on the *Northern Sky* but she was a hardworking ship. There's a saying that a hardworking ship is a happy ship and it's true. In ships that are easygoing with nothing to do the crews get fed up. That's when the nitpicking and arguments start. When everyone is working you don't have time for all that rubbish.

Early in September I came out of the *Northern Sky* to do my RNR (Royal Naval Reserves) training. When I was called up for my training I was told I was to do four weeks at Chatham Barracks. It was September 1956 and the Suez crisis was in full swing. Four of us from Grimsby travelled up to Chatham together. If the Suez crisis wasn't resolved it was highly likely we'd be sent out there as we'd be the latest trained. That wasn't what we wanted to hear! Even so I enjoyed my time at Chatham.

The training was hard, lots of marching and drills. We thought we were fit as we worked long hours and did a lot of hard graft at sea, but it took an officer in his forties to show us just how unfit we really were.

We were lined up outside our barracks waiting for our petty officer to join us when a stunning-looking wren walked by. She looked so smart in her uniform and was very good-looking which set us off whistling at her. Just then our petty officer arrived on the scene and wasn't impressed to say the least. He lined us up outside the offices and gave us a lecture on manners. He told us it was impolite to wolf whistle and marched us smartly over to the arsenal were we were issued a rifle and then marched us all back at the double to the offices. After another lecture on how to behave it was time for him to teach us some manners and he ran us round the whole block at the double with the rifle raised above our heads. He was with us the whole time to make sure that there was no slacking.

The first time round the block we managed quite well. On arriving back at the offices he stood us down. We thought that was it but no sooner were we stood at ease we were off again. We started at 09.00 and he kept us at it until twelve noon. We were in a mess, totally knackered and exhausted yet he was as fresh as a daisy! As we stood at ease the same wren who was the cause of it all walked out of the office. The petty officer asked if anyone had a whistle for her! No volunteers – just silence. 'Good,' he said. 'I'm pleased to see you've learned some manners.' The wren looked round and gave us the cheekiest grin you've ever seen.

During my time at Chatham the warship HMS *Birmingham* arrived at the dockyard for a refit. My brother-in-law Ernie was an engineer on her. We arranged to meet at a pub for a drink the next day and when I arrived he was already there with his shipmates. I got the distinct impression they'd been there for some time. I had a few drinks with them and Ernie invited me to have a look around HMS *Birmingham*. She was massive and I was very impressed but not enough to make me want to join the Royal Navy.

When I was ready to go back to the barracks Ernie said he'd show me the way as he fancied a walk but his mate insisted he take me on his motorbike. His motorbike wasn't much to look at, in fact it was a bit of a wreck, but I'd never been on a motorbike before so was looking forward to it. I got on the pillion and he started the engine and just shot off as if he was leaving the starting gates. I nearly fell off the back as we roared away. I just managed to grab hold of his coat and hang on. I'd only been on the bike a couple of minutes when I realised it was a mistake, this guy was crazy. The bike seemed to have only one speed, full speed. After ten minutes I'd got the impression he'd no clue as to where we were. He slowed the bike down and yelled above the roar of the engine 'I know where we are now'. With that we shot off at full speed again and took the next corner at such an angle I thought we were coming off. There was worse to come. As we tore round the corner and straightened up there in front of us was a huge pile of dockyard rubbish. It was like a small mountain and I hung on like grim death. He swerved the bike violently, missing the main pile of rubbish but running over the bottom edge of it. As we did I felt a coil of wire loop around my foot, ripping my shoe off. It could've been worse though, it could've been my bloody foot!

He slammed the brakes on and I shot forward, falling off. He got off and helped me up. I was a bit bruised and battered and my foot felt sore. I was OK though but he'd scared the hell out of me. He went back to look for my shoe. As I stood there I noticed a building I recognised and realised I wasn't far from the barracks. He came back over with my shoe – the top half of it had been ripped away. I thanked him for the lift but told him I'd walk the rest of the way. Next day I told Ernie about it but he wasn't surprised. Turned out he was a maniac when he was sober let alone when pissed up. If I'd known I'd never have got on in the first place.

Aunty Nancy lived in Erith which was three or four stops away from Chatham on the train. I decided I'd go and visit her and arrived mid-morning. Erith wasn't a big place so I'd no difficulty in finding her. It was lovely to see her and Uncle Albert, they made me very welcome. We had a good old talk about things in general and the family. After tea she asked me to do her a favour and take the dog for a walk as her legs had been playing up.

No sooner had I stepped out the door when the door of the house opposite opened and out came this girl with her dog. I had a girlfriend in Grimsby, but this girl was something else. She was absolutely breathtaking and my first impression was WOW! I asked her if we could walk the dogs together as I didn't know the area very well. She agreed and we got on like a house on fire. I told her I'd be back in Erith in a couple of weeks and we agreed to go out one night. Later I told my aunty about this fabulous girl I'd just met. 'That'd be Pat,' she said with a knowing look. I got the impression the meeting had been engineered. I asked Aunty Nancy if I could come and stay for a week after my training was over. She said it would be fine. Two weeks later on the train from Chatham to Grimsby I told my mates I'd be getting off at Erith.

'What about your girlfriend, you know she'll be at the station waiting for you?' questioned one.

'Just tell her I'll be back in a week or two,' I replied.

It turned out to be three and that was the end of that relationship. Those three weeks were great. Pat was a hairdresser and I'd meet her from work every day and walk her home. We went to the pictures a few times and for walks in the park. Her mother was great, a lovely woman who was always very kind and friendly. Her father on the other hand didn't approve of me at all.

I was now living in Fleetwood, fishing out of Grimsby and courting a girl in Kent! When I came in dock I'd go to the station and catch the first train out, whether it was Kent or Fleetwood, though I did try to keep them even.

On returning to Grimsby I rejoined the *Northern Sky* and stayed with her until Christmas. It was that last trip that we had engine trouble, something to do with the steam tubes in the engine room boilers. On arrival at Harstad we had to go on the slip for repairs.

The engines and generators had to be stopped which meant no heating on the ship. The weather was freezing, lots of snow and ice, so instead of staying on the boat we got some warm clothes on and went ashore. One of the crew was a lad about my age. I think his name was Sid Balldock. We decided to have a walk into town. We could see on the hills behind the town there was a ski slope. We headed off in that direction to have a look and it took us about three-quarters of an hour to get there. I don't know anything about skiing but it looked very professional. The ski runs were marked out with flags and it seemed well organised. We watched for about half an hour when one of the skiers asked if we'd like to have a go. We declined as it looked rather dangerous if you didn't know what you were doing. There was a group of young children just to one side of the ski run playing on homemade sledges; sitting on sheets of plywood, they'd get hold of the front, pull it up, push themselves off and speed away down the slopes. I decided this would be a safer activity for us to have a go at.

There were two young girls of about six years old on skis and very good at it they were too for being so young. Most of the children spoke English so we got chatting and had a go on the sheets of plywood. We spent about three hours with them and had a really enjoyable time. They had some sledges with a seat at the front and handles at the back. You held onto the handles and stood on two runners at the back and off you'd go down the slopes.

I had just got on it and started off down the hill when all the children started shouting 'No, No!' I wondered what all the fuss was about but as I sped down the hill the front of the sledge dug into some soft snow and I was thrown head over heels into the snow. I banged my elbow and for a few minutes wondered if it might be broken. Luckily it wasn't, but it was sore for a few days after.

When we returned to the ship it was early evening, the temperature had plummeted and we were absolutely freezing. Fortunately the ship's repairs were completed and the heating restored.

I did several more trips on the *Sky*. On leaving the river one of the first jobs was to set the log, a length of rope about one and half times the ship's length. The end that is paid over the side has a propeller on it and the other end is fastened to a clock which is fixed onto the handrail on the stern; this records the approximate speed and distance the ship travels through the water. Once at the fishing grounds it's pulled back on board the ship and lashed up ready for the journey home. The trip was hard and by the end of it we were all shattered and pleased to be on our way home. The skipper didn't stop the watches. As it was my watch below they didn't call me out.

As we started for home the skipper instructed the log be put out, but it couldn't be found. Someone said at the start of the trip that it was me who'd pulled in. The mate sent one of the crew to wake me up. I was fast asleep, dead to the world, and not happy about being woken up. I told them it wasn't me that'd pulled the log in and that I'd no idea where it was. They didn't believe me and came back several more times to ask me again. In the end I politely told them to 'F★★k off' and was later politely told I'd got the sack. I was sorry as I'd enjoyed my time on the *Northern Sky*. I wouldn't have minded so much if it had been me that pulled that damn log in.

After I left I did a couple of trips in the *Northern Prince* and one in the *Northern Sun*. They were run-of-the-mill trips, nothing exciting about them. I then joined the *Northern Sky*. Her skipper was Vic Meech; the last time I'd sailed with him he'd given me the sack for oversleeping and not getting out on deck on time, so I guess he must have forgiven me by then!

CHAPTER 15

LOSS OF THE LIBERTY SHIP *PELAGIA*, 1956

The SS *Pelagia* was a vessel of 7,238 gross tons, 442ft long with a beam of 57ft. Built in 1943 she was owned by Eastern Seaways Corp. as the SS *Sea World*, a steel-hulled American freight vessel of the liberty type. The master was John M. Matandos from Houston, Texas.

The *Pelagia* left Norfolk, Virginia, on 17 August 1956 bound for Antwerp in Belgium, arriving at her destination on 4 September to discharge her cargo of coal. She was inspected in Antwerp and no major faults were found apart from some damage caused by the buckets when unloading the coal and these where repaired. She left Antwerp bound for Narvik in Norway where she would load 10,190 tons of iron ore destined for the Bethlem Steel Company, Philadelphia. She arrived in Narvik on 12 September.

On 14 September loading was complete. The ore was loaded into five holds and none of it was trimmed by hand. It was peaked in the middle of each hold by the movement of the loading chutes. The vessel's hatches were secured with hatch boards and tarpaulins, inspected and found to be satisfactory.

The *Pelagia* departed Narvik at around 13.00 on 14 September. The weather was good with a moderate south-east wind and a calm sea. During the night, however, the weather became worse. At 05.00 on 15 September Harstad radio (Norway) radioed the weather forecast to the *Pelagia*. By now the weather in the area had deteriorated with wind force 9 WSW and a heavy sea running. Around 05.10, while steering a course of approximately 260T, a heavy sea struck her, awakening most of the crew. This sea demolished the wooden and canvas dodger of the flying bridge.

At 05.30 the chief officer called all hands on deck to start removing the tarpaulin off No.4 hatch intending to place it on No.1 hatch. Before this was done, the deck force received orders to re-secure the No.4 hatch again due to heavy seas forward. When the work party had re-secured No.4 hatch they were sent up to the boat deck, starboard side, where 1 and 3 lifeboats were found to have been pushed back off their cradles by the force of the sea. Both boats were jammed into their after davits. After lashing 1 and 3 boats, the work party went to boat 4 which had been partly lifted from its cradle. The boat was lashed in place to prevent it swinging.

External evidence of damage to the boat's hull involved dents but no punctures. At about 06.00 the tarpaulins, three in number, on No.1 hatch had been torn away, and approximately four hatch boards were missing. The tarpaulins and hatch boards weren't replaced due to heavy seas breaking over the fore part of the vessel.

Around 11.30 the vessel's steward was told by the chief officer the vessel was being held on course, into the seas.

At 12.45 an SOS message from the *Pelagia* was received by Norwegian radio stations Rørvik, Tromsø and Harstad, giving a position of 66 30N 10 30 East.

At 12.45 Rørvik radio received the following message: 'SOS full of water cannot keep water out need immediate assistance position now 17 15N 08 35E at 1100 GMT since steering course 260 degrees QRA Pelagia from Narvik bound Baltimore.'

At 13.00 the general alarm was sounded and the crew, in lifejackets, mustered at their lifeboat stations. The steward observed black water coming out over the combing and being regurgitated from No.1 hold. Under the direction of the boatswain, No.2 lifeboat was uncovered and swung out. No.4 lifeboat was partly uncovered when orders from a person unknown stopped the operation. The boatswain directed one able seaman, one ordinary seaman, one steward, one 2nd cook, a baker and a mess man to enter No.2 lifeboat.

At 13.50 Tromsø advised the Rescue Operations Centre at Bodø, Norway, of the SOS and they alerted aircraft and surface vessels.

The six men on No.2 lifeboat were lowered into the water, riding on a sea painter attached by the chief officer to the forward deckhouse stanchion, port side. The lifeboat rode on the painter until approximately 13.20, when it broke and the boat drifted aft. All but two oars had been broken trying to fend the boat off from the ship. The boat cleared the *Pelagia*'s screw which was still turning at 40–50rpm. Men on deck told those in the boat that land was reported some 5 miles away.

At the time the lifeboat went adrift the *Pelagia* was upright but down by the head with seas breaking over nos 1 and 2 hatches and as far aft as No.3 hatch. No panic or apparent distress was displayed by the personnel that remained on board.

At 13.13, Rørvik radio received a corrected position of 67° 15N and 11° 35E, course 260°T. At 13.21 the station received another message indicating the lifeboat adrift. No further contact was made with the vessel.

Around 14.20, Bodø Rescue Operations Centre received a message reading 'now sinking', the signal thereafter fading. The German boat MV *Kelkheim* reported intercepting the message of sinking around 14.35.

During the night the lifeboat remained adrift. The lights of various vessels were observed and attempts to signal them with the lifeboat flare pistol were made. Some five flares, although propelled by the cartridge charge, failed to ignite, the first seven flares fired properly and the parachute flares ignited. After exhausting the parachute flare signals, there were attempts amde to use the handheld flares. It was believed that to ignite these flares matches were needed, and the survivors endeavoured to use the matches stowed in the boat. However, these, together with the flashlight, were found to be wet when the container was opened, the container itself showing signs of prior interior wetting.

At dawn the men in the lifeboat discovered that one of their number had died. The temperature of the sea had remained at 8°C. The boat had been taking water over the gunwales and through an apparent leak which had remained undiscovered by the seamen. The sail, which they'd used as a cover during the night, was hoisted, upside down, during the early daylight hours as a mark of respect.

Around 07.30 the lifeboat was spotted by a Catalina aircraft of the Norwegian Air Force. The aircraft signalled the MV *Kelkheim* and the British trawler *Northern Duke*, both in the area in which the boat lay, and they headed towards the lifeboat.

Around 07.50, the MV *Kelkheim* sighted the lifeboat and manoeuvred to fire lines with her rocket equipment. The *Northern Duke* manoeuvred across the MV *Kelkheim*'s bow, came alongside the lifeboat, and took the five survivors aboard. The wind was still at force 7, WNW, the sea WNW was moderating.

The *Northern Duke* took the lifeboat containing the dead man aboard with her net-handling gear and proceeded to Harstad, Norway.

The entire procedure from the *Northern Duke* crossing the MV *Kelkheim*'s bow until the lifeboat was brought aboard took only 10 minutes. At 18.30 the MV *Kelkheim*, despite heavy seas and force 9 to 10 wind, had reached the location of the ship but found nothing.

On intercepting traffic during the night regarding sightings of lights or signals, the MV *Kelkheim* headed southwards at varying speeds. At about 04.30 on 16 September the MV *Kelkheim* sighted and recovered two hatch boards, a ladder and a lifering buoy marked 'SS *Pelagia* – New York'. Around the same time the vessel received a report of oil slicks sighted about 5 miles WSW of her 04.30 position.

The master of the MV *Kelkheim*, assuming the lifeboat would move more rapidly before the wind, headed ESE for about two and a half hours without sighting anything before turning his vessel on a NW heading. Around 07.45 they sighted and retrieved a dented lifeboat air tank marked *Pelagia*. While engaged in picking up the tank she was signalled by the Catalina aircraft to precede towards the lifeboat's indicated position.

The *Northern Duke* arrived at Harstad, Norway, on 17 September, where the survivors were sent for medical examination. The body was shipped from Harstad on 21 September to Bergen, Norway; from there it was sent to Playa De Fajardo, Puerto Rico, for internment.

On 17 September, the Norwegian destroyer *Trondheim* picked up the body of a seaman wearing a lifejacket marked *Pelagia*. The body was later identified as that of an able-bodied seaman from the SS *Pelagia*. The lifeboat put ashore by the *Northern Duke* at Harstad, Norway, was kept under police custody.

There were thirty-seven seamen on board the *Pelagia* when she departed from Narvik, Norway. A total of thirty seamen including Captain Matandos were missing without trace after she sank. Only two bodies were recovered and five men rescued.

The master of the *Northern Duke* was Captain J.W. Greene – my dad – and the captain of the MV *Kelkheim* was Kurt Kreiger. Both had showed great skill and seamanship in their actions leading up to the finding and rescue of the survivors in the *Pelagia's* drifting lifeboat.

Statement made by J.W. Greene, master of the ST Northern Duke

We sailed from Grimsby at 10.45 GMT, 12 September, 1956. On Saturday 15th Sept, 1956, at 16.27 GMT, we received a distress message via Wick radio from the American S.S. Pelagia stating that she was leaking and needed immediate assistance in pos Lat 600 15' north Long 110 35' East, also that she had one lifeboat adrift with four persons on board.

Aircraft could not leave Bodøe airport on account of the bad weather, also three ships estimated arriving at S.S. Pelagia's pos at – 20.00 G.M.T. S.S. Pelagia transmitting now sinking, end of message.

At this time my pos 120 miles to the SW [of] S.S. Pelagia's pos. I was steaming at ordinary full speed, the wind was WNW force 8, seas 8. And we were taking heavy seas on the port side; I had been up all night on account of the weather. We altered course towards the S.S. Pelagia's position and altered the ship to the facts. In view of the fact that vessels were due from 20.00 GMT onwards and considering the distance we were away I did not think that we would arrive in time to be of any assistance.

We kept a continuous W/T and R/T watch from the first ships arriving; we listened to conversations between ships and shore stations, and also plotted the positions given by the searching ships. We closely followed the search and plotted on the chart the area that was being searched.

The ships concerned from about 20.00 GMT onwards were the ST Northern Sceptre, British Navigator, the German vessel call sign DLBE, and an unknown Dutch vessel. The search commenced from the S.S. Pelagia's given position to windward to the SE to 10 miles SW, with no results. Then the ships commenced searching in a NE direction in towards west fjord with no results. I could not see the Pelagia being to the NE because she would have had a lee from the land, and not have been taking heavy seas. Also the absence of oil and wreckage was puzzling the search ships.

At 22.05 I realised that there was more than a possibility that the Pelagia had been more to the Westward or SW of the given position, at least I saw no point in proceeding to the area which was being searched with negative results. I informed the Chief Engineer to take the block out of the expansion and to proceed at utmost speed, with six lookouts not including myself and the mate.

At 23.55 GMT we crossed the NE edge of Trean Gully in 200 fathoms of water, in position 30 miles to the SW of the S.S. Pelagia's last known position, and commenced searching in a NE direction. I then steamed 20 miles until I was 10 miles SW of the Pelagia's position then turned SE at a reduced speed and back to the SW again. I continued searching on SW and NE courses, always turning downwind with long sweeps to the SE and keeping well to the SW of the searching ships.

At 05.10 GMT 16 September, whilst on a SE sweep, we ran across a large area of oil and wreckage. We reported this to Rorvick radio and the searching ships. The S.S. British Navigator and German merchantman joined me. We passed a bright yellow-painted oar but on coming round I could not locate it again for the sea running. We also passed a large amount of white-painted wood in this position. I got the impression that the Pelagia had broken in two. We then passed several bed mattresses and later recovered a lifejacket marked Pelagia.

We continued searching but could discover nothing more than wreckage. The wind had decreased to about 5 or 6 but with a good sea running. I continued searching to the SE until I ran clear of the wreckage and continued at slow speed for some time still going SE.

At 07.00 the Grimsby trawler ST Northern Sceptre steamed passed me and he radioed he was resuming his course and speed. I decided that I was far enough to the SW and with the tide being stronger to the NE than the SW I thought that I would try more to the NE as I was now well to the SE of the wreckage. About this time the helmsman, J. True, reported having observed something to port; we altered course to port and came onto a NE course and I increased to half speed. As we were trying to locate the object that had been sited, an aircraft which had been searching began making circular turns towards me, and turning to starboard to the Eastward. I knew he had seen something and altered my course to East and increased to full speed at 07.45 GMT. The German merchantman was also steaming round and came on to a NNE course.

After a few minutes I observed the tip of the mast of the lifeboat. The aircraft dropped a smoke flare near to it. Shortly afterwards I made out the lifeboat and went alongside.

The German steamer and I arrived at the same time and he crossed between me and the lifeboat passing it about half a mile to the NW and overran it by about a mile. I pulled to the southward across the German's stern and then came hard to port to bring the wind about 2 points on my port bow and the lifeboat under the lee of my starboard bow.

The German steamer was now laid broadside to the wind and was rolling very heavily. He had thrown his scrambling nets over the port side and was firing rocket lines that did not reach the lifeboat. I was by this time closing in on the lifeboat. The mate dropped a heaving line into the lifeboat from off the starboard bow; one of the survivors made the line fast and as I was going astern the lifeboat had pulled along my starboard side aft of the rigging.

The mate jumped into the boat to help the survivors on board, we had also secured the boat by the stern. At 08.25 the first survivor came on board followed by the other four. We got the dead man on board and the mate shackled the fore Gilson into the bow and the amidships Gilson into the stern and we hove the lifeboat aboard forwards of the winch. At 08.30 we wired Bodoe radio and the British Navigator that we had picked up the lifeboat and survivors.

Summing up the situation, we were making quite a lot of water into the engine room through the engine room casing; also the last weather forecast we had received was wind WNW force NINE. In view of this forecast and not being certain to the extent of my damage I immediately proceeded at full speed for the lee of the Lofoten Islands.

The five survivors seemed to be in very good condition. I instructed the mate to organise dry clothes and towels and plenty of massaging E.C.T. for them; they had some hot drinks and something to eat then all turned in for some sleep.

As the survivors required no immediate medical attention I decided to proceed to Harstad where I could also get repairs done. Bodoe radio asked me if I would land the survivors at Bodoe. I replied that I was proceeding to Harstad. Bodoe informed me that he had a link call for me from Flying

Control. Flying Control told me to come into Bodoe. I asked who Flying Control was but they did not reply to my question. I had no intentions of entering the south fjords along this coast with the wind and heavy seas running from the WNW. If I did then it would only be in an extreme emergency and I would require a pilot, and I could not see a pilot boat coming out in that weather. I informed Flying Control that I would be willing to land them in a more sheltered place if they wished and to make me an official message to that effect. Flying Control told me to carry on to Harstad. We arrived at Harstad at 23.45 GMT on Sunday 16 September, 1956 and landed the survivors. The lifeboat that the survivors were rescued from was on the dockside at Grimsby for quite a long time after that, what eventually happened to it I have no idea.

(Information U.S.C.G./Grimsby Mutual Insurance)

For the rescue of the crew of the *Pelagia* my father was presented with a gold watch from the President of the United States, who at the time was President Eisenhower.

CHAPTER 16

ROYAL LINCS, SEPTEMBER 1957

Northern Trawlers had taken over John Butt's fishing company and *Royal Lincs* had belonged to the Butt Group. She was a large modern trawler built in Germany with excellent accommodation and all mod cons but that's as far as it went. I've sailed in good ships and bad ships and she was one of the bad ships. The skipper was one of the worst I ever sailed with, he was a right bastard!

We sailed for the White Sea. The passage down was very bad so we picked up the pilots at Lodigen, sailed through the Norwegian fjords, dropped them off at Honningsvåg, and turned east for the White Sea plaice grounds.

When we arrived at Cape Cannin and Novaya Zemlya the weather was atrocious with gale-force winds and heavy icing. The skipper eased the speed down and ordered the mate to call all hands out. This was the last time we saw our bunks again except for the watch below.

We spent six hours chopping away the ice before we could get the fishing gear over the side and start fishing. The rest of the trip was eighteen hours on and six off every day as the skipper wouldn't give us a break. We were fishing for plaice, and they took a lot of shifting. On most hauls we'd have enough fish to keep gutting until the next haul.

We had one break during the whole trip, and then only because the weather took a turn for the worse and we had to get the gear on board and dodge.

I was on the bridge watch with the mate. The skipper was there too when the radio operator, who was the skipper's son-in-law, came up. Skipper looked at him and barked, 'What the bloody hell is up with you. You look like s★★t!' The radio operator was full of flu and felt bloody lousy. 'I've got just the cure for you,' said skipper. With that he disappeared into the chartroom and returned with an electric light bulb. He gave it to the operator and instructed him to go up the foremast and change the bulb in the masthead light.

I couldn't believe it, I thought he was joking. I stood looking out the bridge window watching as he climbed up the rigging (which was coated in ice) to the top of the mast. My heart was in my mouth hoping he wouldn't slip and fall. When he finally came down from the mast and onto the bridge he was absolutely frozen through. His face was blue and he'd a coating of ice on his eyebrows. The mate turned to the skipper and suggested a dram of rum would be a good idea. 'He can f★★k off, he's not getting one.' What a rotten bastard. I felt really sorry for him. Had the skipper told me to do that in such treacherous weather I'd have told him what to do with his masthead light and his bulb.

Credit where it's due, though, the skipper was a very good fisherman and one of the port's top earners, but the way he treated his crew was diabolical. He had a regular crew and I'm sure most were afraid of him. It was only the money that kept them there. The third hand seemed to get it most, though; no matter what he did it was wrong. Every time he hauled the gear the skipper would be screaming and shouting at him and had him running up and down the deck like a scalded cat. It didn't matter what he was doing or where he was, he'd be in the wrong place and doing the wrong thing. We called it being spot balled but really it was bullying. He made that man's life a misery.

On the second day we badly ripped the trawl. We were told to chop it off and put a new trawl alongside. While changing the net one of the crew said, 'That's it Jim we'll be on deck for the rest of the trip.' 'You'll see,' he replied simply when I questioned what he meant. As soon as the fish was gutted we started repairing the net now on the port side. We made more progress after each haul and by the end of four days' fishing it'd been repaired. The next haul it had small holes in it. The skipper said we had to go through it again. These nets are 250ft long so it takes some time to go from one end to the other repairing every little hole. When we'd finished I thought thank God for that.

The weather was still freezing and we had to keep the ice down. After each haul, when the fish had been gutted and put away, we'd chop away the ice and shovel it overboard using hatch batons, axes or whatever else we could find. The entire time skipper would be watching and as soon as we finished and started leaving the deck he'd ring down to the engine room to haul in the net. As we were working in shoal water it only took five or six minutes to heave the gear up. No chance to go aft for a pot of tea or warm up. The fishing was good most of the trip and sometimes we'd still have fish on the deck when it was time to haul the net again. When we went back into the fish it was frozen solid but we still had to gut them. Between chopping the ice away, gutting the fish and mending the trawl, we were hard at it eighteen hours a day and with the weather freezing and blowing all the time it was no picnic.

One haul we had a hole in the cod ends so we had no fish at all. When we'd mended the cod ends and the net was back on the seabed, there was no ice left to chop, no fish to gut and no net to mend. I thought we'd get a couple of hours off. Who I was kidding? As we went aft skipper called the mate to the bridge. We'd just got our gear off and settled in the mess deck with a pot of tea when the mate came and informed us the skipper wanted to make some changes to the net we'd just spent four days repairing. With that the ship's whistle started blowing, a signal that the skipper wanted us on deck. Everybody dashed out leaving mugs of tea half drunk. I decided to finish my tea first and when I got on deck the mate gave me a right telling off.

On our last day fishing we were still messing about with the net which by now had had so many changes made it was in pristine condition. Around 10.00 the mate told us the skipper had decided the net wasn't really worth keeping, he wanted us to strip it down and save all the good bits and pieces. The rest was rolled up into a big bundle and dumped overboard. What a bastard! I did three trips and after the third I signed off. The ship's runner, Bill Robinson, called me to the office wanting to know why I'd signed off. He said a lot of people would give their right arm to be on the ship and earning that sort of money. I told him about the conditions and what the skipper was like and that I wouldn't work with a skipper like him for any price as he went out of his way to make the crew's life a misery. Bill said he'd give me a ring when he had another ship for me. He wasn't very pleased with me and I gathered by the tone of his voice that I'd be having a long walkabout. When you'd been a bad boy they'd keep you out of work for two or three months to teach you a lesson. This often happened in the deep trawlers.

CHAPTER 17

EQUERRY, 1957

After leaving the *Royal Lincs* I knew that I'd be at home for a spell. It didn't bother me as I'd some money saved up. I stayed at home and enjoyed my time ashore. I'd been home for about seven weeks when I got a call from the office. It was Bill Robinson, the ship's runner, asking if I was ready to go back to sea, this time on the *Equerry*. I said I was and he told me not to bother coming down to the office as he'd sign me on the ship the morning we sailed.

When I arrived ready for sailing I couldn't find the ship. I walked up and down the North Wall looking at all the big deep-water trawlers but there was no *Equerry*. I'd never seen the ship before but I knew she was one of the trawlers bought by Northern Trawlers from the Butt group. When Bill arrived on the North Wall I asked him where she was and he gestured for me to follow him. We walked along the North Wall and there amongst all the old North Sea ships was the *Equerry*. What an old wreck and absolute rust bucket she was! I took one look at her and told Bill that I'd only be doing the one trip. 'That's OK,' he said. 'Come to the office when you land.'

There was nothing about this ship that gave me any confidence. I asked Bill if we were going to Iceland; he grinned and told me we were going to the Faroe Islands. 'I doubt she'd even make it to Iceland,' he joked. Great, I thought.

The *Equerry* was a coal-burning trawler built in 1929. She was 170 tons and about 134ft long with accommodation very much like the old *St Philip* in the forefend of the ship.

The skipper was one of the best. A really great skipper to sail with, he worked with the crew not against them, which made a big change from the last ship. I don't remember the mate's name but that didn't bother me as I didn't like him anyway. Most of the crew were young except for the third hand. The word misery was invented for him as I've never met such a miserable sod in my life. As time went on and I got to know him better I found he wasn't too bad really, it was just the way he was. I have no recollection of the engineers or the cook.

On the way to the Faroe Islands we had a couple of stops for engine repairs. This turned out to be a regular thing as she was always breaking down with one thing or another. We had a good crew and we got on well together. She was a happy ship, there was no rushing and everything went smoothly. As we approached the Faroe Islands we were told we'd be going straight into a little fishing village on the east coast of the island, as the engineers had a job to do on the engine.

The *Equerry* fished the Faroes on a regular basis. It turned out she was a regular visitor to the harbour and as we arrived some locals would be down to take our ropes and help us tie up. I don't think the job in the engine room was very big, more an excuse for a night on the booze with the locals. A couple of the crew even had girlfriends there.

The *Equerry* turned out to be a good ship despite her age and the numerous breakdowns and I stayed with her for several months. One trip we went into harbour and most of the crew went ashore. Two lads about my age, the skipper and third hand, stayed on board. The two lads and I were fooling around and making a lot of noise. The skipper came out of his berth and gave us half a bottle of rum between us and four cans of beer each. He told us to bugger off ashore with the rest of the crew so he could get some sleep. It was early evening and as it was summertime it was still light. We sat on the fore hatch and had a couple of drams and a can of beer each.

One of the lads suggested we go for a walk into the mountains and we all agreed and set off. We followed the road from the village and into the hills. As we walked along we passed several cottages with what looked like joints of meat hanging up outside to dry; it was pilot whale meat. Pilot is a small whale which arrives in the Faroe Islands each year. Years ago islanders would go out in their boats and drive the whales down a very narrow fjord until they were forced onto the beach. Others waited for them, men, women and children with anything they could use for a weapon. As the whales hit the beach they killed as many as they could. The meat was hung up to dry and used for food in the winter months.

As there was nobody about we decided we'd try some, so we cut a piece off for ourselves. We walked down the road eating this whale meat. It was very tough like an old boot and not very nice. A little further along we came across some chicken hutches and decided to get some eggs for the cook. I kept watch as the other two went into the hutches to get the eggs. There were only about half a dozen so we carried them in our caps. As I was walking along something hard hit me on the back of the head. I thought one of my mates had thrown a stone at me so I turned round and called them all the rotten sods I could think of. It wasn't a stone my mate had thrown but one of the fake eggs used to encourage the hens to lay.

We sat by the roadside and drank our last can of beer. The scenery was breathtaking – we could see all the way down the mountain to the harbour nestling at the water's edge. It was an impressive sight. The evening was peaceful and calm. If I'd been on my own I could've happily spent a couple of hours just enjoying the peace and quiet. Instead I was with my noisy mates who were discussing the idea of getting a couple of chickens to take back to the ship. I wasn't keen on the idea but was outvoted two to one. When we went into the hutches the birds made so much noise it was unbelievable but we managed to get two chickens and put them in a sack then made our way back down the mountain back to the ship.

By now it was getting dark, all the lights in the village and the harbour were on, which looked great from halfway up the mountain, but it was a hell of a long way to walk. By the time we got back it was well and truly dark. Most crew were already on board and turned in. As we entered the village we came across a bicycle laid against a wall. My mate being fed up with walking said he'd go back on the bike. The road was downhill all the way. We watched him peddle down the hill, picking up speed as he went. It wasn't until he got to the edge of the quay that he realised there were no brakes and the only thing he could do was jump off the bike. He finished up in a heap on the quayside and the bicycle disappeared into the water.

There was nobody about and the ship was quiet so we went into the mess deck and made a pot of tea and some sandwiches. As we were tucking into the sandwiches we chatted about the best place to keep the chickens. We decided the anchor chain locker would be best as it was cool down there. After we put the chickens in the chain locker we turned in for a good night's sleep.

We were called out at 05.00 the next morning. During the early hours a ship from Aberdeen had arrived in the harbour. There were no signs of life on her so the crew must've all been turned in. We had breakfast then the skipper gave the order to let go of the ropes. The mate and I went on to the bow to let go the forward rope, the others went aft and one man went on the quay to throw the ropes off the quayside bollards.

I noticed the harbour master making his way down to the ship; the mate was first onto the bow. He looked over the side and all hell broke loose. He started screaming and shouting at the man on the quayside to throw the ropes off as fast as he could. What's got up his nose, I thought? I looked over the bow and there hanging on the rope was the bicycle that had gone in the night before. Now we're for it I thought, he'll give us a dog's life when we get to sea.

The bow rope was slipped. As it went into the water the bicycle slid down and disappeared into the water again. The harbour master had just reached the edge of the quay but didn't see the bike. The mate wished a cheery good morning to him. In return he wished us good fishing. As the boat slid slowly away from the quay under his breath the mate whispered, 'Just wait until we start fishing'.

An hour later we stopped, put our nets over the side and started fishing. The skipper wasn't best pleased with our exploits the previous night. Our punishment was to check the net store and everything down there. We had to make a list of everything in the store and then put it all back so we were knew exactly where each item was.

Our first haul was a good one so the skipper calmed down a bit and things returned to normal, but not for long. We'd been gutting fish for about an hour when the skipper called the three of us on the bridge. He said he'd just spoken to the skipper of an Aberdeen trawler who'd informed him his ship was under arrest. Apparently two chickens and a bicycle had gone missing. The police were searching his ship and they wouldn't be allowed to leave until the property had been recovered. He asked if it was anything to do with us, which of course we denied. My mate said it would be the 'Deenies' as they were all robbing bastards and we managed to convince him we'd nothing to do with it.

Fishing didn't stay good for long. After two or three hauls it had disappeared. The skipper decided to change fishing grounds, which meant a two-hour steam. I had to take the wheel and steer the ship for the first hour then my mate John was going to relieve me. After my hour was up the bridge door opened and in he walked. It was unfortunate the skipper was still on the bridge. As he entered I turned round and looked at him and burst out laughing. 'What's so funny?' my mate asked. The skipper turned round to see what I was laughing at. He didn't say anything but he didn't look happy. My mate took over the wheel and I made a swift exit for the door. 'And where do you think you're going, get back here now,' the skipper growled. John was steering the ship with a big daft grin on his face wondering what was going on. The skipper asked him again if he knew anything about the chickens.

'Not that again, honestly skipper I don't know a thing about them.'

Any minute now, I thought, this is it.

The skipper paced the bridge then went back to him. 'Last chance,' he said. 'Are you sure you don't know anything about those chickens?'

'I don't skipper honest.'

I thought skipper was going to thump him. I was trying my best to keep a straight face but I was cracking up inside, it was so funny.

'Then can you explain this?' said skipper and took my mate's hat off his head and shoved it under his nose. John looked at his hat; it was covered in chicken feathers. That was it, I burst into fits of laughter. He'd been down the chain locker plucking the chickens. He'd checked himself for any sign of feathers before coming up to the bridge but unfortunately he hadn't thought of looking at his hat.

The chickens were very tasty. We made a big pan of stew on the stove down the forecastle and gave the skipper his share. We daren't do anything else! His only complaint was it could've done with a bit more salt in it. The rest of the trip went OK. We had a few good laughs about those chickens and eventually the skipper laughed too.

When we arrived in the River Humber we were instructed to drop anchor and wait for the lock gates to open. As we dropped the anchor chicken feathers went flying out with it. It looked like a snowstorm with feathers everywhere. The ship anchored nearest to us complained his deck was full of feathers and he had to get his crew to wash the decks down before docking.

I did one more trip in the *Equerry*. On our way down to the fishing grounds we had to spend three days in Aberdeen with boiler trouble. The job was patched up as best as they could, but was only a temporary repair; once we got back to Grimsby they'd have to decide whether to spend money repairing the ship or send her to the scrap yard. Altogether that trip wasn't very good although we didn't go to the Faroe Islands and therefore didn't have to face the music about the chickens.

The day after we left Aberdeen we discovered we had stowaways on the ship: five of the biggest rats I've ever seen. They were first sighted in the early hours of the morning foraging about on the foredeck looking for bits of food. I can honestly say in all my time on trawlers I hardly saw any rats at all. Usually the rats on board a ship were of the two-legged variety, and I sailed with quite a few of those.

The biggest part of the trip was spent trying to catch or kill them. We tried all sorts; we put food out on the foredeck and waited to ambush them. We were armed with hammers and spanners and anything else that we could throw at them, we even tried making traps. But they were having none it. They were the craftiest bunch of rats you could imagine. One day we caught them halfway along the starboard side of the ship and chased them into the liver house. All five of them jumped into an old liver boiler, we slammed the lid and screwed it down tight.

In the liver house there were three boilers used to boil the livers. Now we had five rats trapped in one. The question now was what we were going to do about it. One suggestion was to turn on the steam but we thought this was a bit cruel so I suggested turning on the water and drowning them; we had a unanimous decision on this one. We filled the boiler with water then went away and left them for six hours. After a good six hours four of us went back and started to unscrew the clamps on the boiler lid. These lids work by a heavy weight, when you've undone the clamps the lid automatically flies open. We wanted to see if the rats were floating on the water, so we crowded round the lid trying to get a look inside. The lid flew up, and out jumped the fattest five rats you'd ever seen, they were blown up like balloons! The next minute we were all fighting to get out of a doorway made for one person to get through, they absolutely terrified us. The rats disappeared along the deck, back to wherever they'd been hiding. After that we left them alone. The rats had won.

When we arrived back in Grimsby I decided I'd stop ashore and get my third hand's ticket. I went to the nautical school in Orwell Street on the docks to sit for it. There I met a lad the same age as myself called Alf Hodson, whose father was tutor at the school. We became good friends and sailed together on several occasions after. We both passed with flying colours on 12 December 1958, aged nineteen.

CHAPTER 18

BOMBARDIER, JANUARY 1958

I signed on the *Bombardier* after Christmas and we sailed for the White Sea. Vic Meech was skipper and as this was my third time with him I knew what to expect. I was pleasantly surprised to find he'd really changed. He was more easygoing than he used to be, but not too much. He was now an established skipper. People don't realise how hard it is when you first go skipper. Usually the first trip as skipper is a nightmare.

The *Bombardier* was a nice enough ship but boy was she a terrible sea ship. She'd pile the seas aboard when you least expected and I was washed along her decks many times. The skipper told me he thought he'd been promoted when they offered him the *Bombardier*. She was a much more modern ship than the *Northern Sky*. But that's where it stopped; she must have been a nightmare to be skipper on. One example of her sea-keeping qualities was when working the Norwegian coast. The weather was gale force 8 but not bad enough to stop fishing and other trawlers around us were still fishing. Twice that morning we'd washed the fish over the side while we were shooting our nets to the seabed. It was fortunate nobody was in the fish at the time. On both occasions a large sea smashed over the rail and washed the fish back overboard. Of course the crew were having their usual moan about the skipper being a stormy bastard and how we shouldn't be fishing in weather like this. On the third haul we hit the jackpot with five bags of cod and haddock, all excellent quality. There were approximately 150 10-stone boxes. With the as weather bad as it was we had a struggle getting the fish on board and took several big seas in the process. When it was on board the skipper told us to get all the gear in as fast as we could and get the ship's head to wind. He had no intention of losing such a big haul.

At twelve noon the gear was on board and the ship's head was to wind. The skipper sent everybody to dinner, leaving the third hand and decky learner gutting. As we were halfway through our dinner we heard a large sea crashing aboard. The ship lurched to starboard and gave a great big shudder. We carried on with dinner as this wasn't unusual in such bad weather. The next thing the third hand and decky learner, shaking with fright and as white as ghosts, were stood in the alleyway absolutely wet through. The third hand told the skipper all the fish had been washed over the side and there wasn't one fish left on the deck. 'It's taken the lot and if I hadn't hung on, it would've taken me and the decky learner with it,' he reported.

The decky learner told us later that when the ship took the sea it was huge. It smashed them down to the deck and he hit his head. He was in a dazed and shocked condition. He didn't know what was happening and thought that he was going to be washed over the side. He felt the third hand, a big strong chap, grab hold of him round his waist. The third hand hung onto him and the trawl warps until the water and fish had cleared the deck then helped him get safely aft.

The skipper of a ship who was alongside us when we took the sea told our skipper we were laid so far over to starboard all he could see of our ship was the hull, he honestly thought we'd rolled over.

One trip while fishing on the Norwegian coast most of the crew was struck down with flu. It got so bad we had to go into Harstad for medical treatment. The doctor examined us and fortunately no one needed hospitalisation. The doctor gave out a flu mixture and injections. He advised we stay in harbour for another twenty-four hours, and said he'd come back before we sailed to check on us. Within twenty-four hours there was a big improvement in the health of the crew and we were allowed to continue on our trip. The doctor gave the skipper some injections and told him to give the worst affected an injection in the backside once a day. The skipper passed this job to the third hand so we christened him Dr Death. I was glad I didn't need them as there's no way I'd have let him anywhere near me with a hypodermic syringe.

During the time we'd been alongside some of the crew had been on the booze, drinking cheap brandy rather like firewater. We left Harstad with some still under the influence of the booze. The worst were Harry and George. Harry was being a nuisance and kept offering George to go on deck and fight. George didn't want to but told Harry he could beat him with one hand tied behind his back. Harry insisted and kept coming back into the mess deck pestering him. George was a very fit, strong man; he was also as hard as nails and knew how to look after himself. Harry wasn't really built for fighting; it was just the drink talking. George finally gave in and they went on deck. I wouldn't really call it a fight as George only hit Harry once and he was out like a light. As he fell to the deck he hit his head on the ship's rail, sustaining a gash just above his eye. George went back to the mess deck and carried on as if nothing had happened.

About ten minutes later one of the crew came into the mess deck and asked the third hand to take a look at Harry, who was still laid out with one eye closed and bleeding from a cut above it. Harry needed stitching up. The skipper took a look and instructed the crew to take him down to the mess deck and that we'd be taking the ship back into Harstad to get Harry's wound stitched up. The third hand asked the skipper to do it, but he refused so the third hand said he'd do it. After all, we had the gear for the job in the medical chest.

Four of us took Harry down to the cabin. We laid him on his back and two sat on his legs and two more held his arms down. The third hand knelt across his chest ready to stitch his eye. Harry was still half pissed so he wasn't really aware what was happening to him – yet. The third hand was a stocky chap with hands like shovels and fingers like pork sausages. Not ideal for the delicate job of stitching!

When he was ready, one of the crew broke open the container holding the needle and passed it to him. He grabbed both edges of the wound with a finger and thumb and squeezed them together with the finesse of a butcher slaughtering a lamb. He stuck the needle into the wound and pulled it out the other side. Harry woke up quick and started struggling. The third hand told him to keep still or he'd thump him. Harry was rapidly sobering up and decided it best to do as he was told. The third hand tied off the stitch and did a second. Harry was now truly sober and insisted two stitches would be plenty. The third hand said it really needed another but Harry flatly refused.

The skipper was pleased with the job and gave the third hand a tot of rum and poured one out for Harry. The third hand took the dram and drank it himself; Harry had had enough booze for one day. The stitching job was well done and it healed up nicely.

At the end of March we stopped fishing the Norwegian coast and started fishing at Iceland. On the first trip we got foul of the gunboat. We were with a couple of ships fishing inside the Icelandic territorial limit line. The weather was bad, the wind blowing from the north-west force 7 with a high swell. We'd only been towing our fishing gear for about an hour when the Icelandic gunboat *Aegir* arrived on the scene.

The skipper sounded the alarm to get the gear up as fast as possible. It was my turn to heave up. I got the winch going and was heaving at full speed. Other ships were doing the same. I watched as the gunboat steamed full speed towards us. He passed two ships on the way and I wondered what he was up to, they usually grab the first ship they come to. He must have had someone in mind. As he steamed down our port side I could see the crew uncovering the gun on her bow. Now level with our winch he pointed his gun at us and fired. I left the winch and ran to the starboard side to get out the way – there was no way I was going to stand there with him firing shells at us. As I got round the starboard side the mate was making his way smartly to the winch which was still going full speed. The gunboat was now alongside us and I could hear him shouting to the skipper on the loud hailer, then he fired another shot.

The skipper shouted down to the mate to ease the winch down. We weren't going to get away from the gunboat. We got our gear on board. The gunboat was sending across his inflatable boat. Our skipper was invited on board, not that he had any option, where they would discuss positions.

The inflatable wasn't very big, and powered by an outboard motor. They had trouble getting alongside our ship due to the heavy swell, the inflatable was riding up and down and the skipper had great difficulty getting on board. He was taken to the gunboat where the two skippers checked positions. They both agreed he was inside the limits. The gunboat captain told skipper we'd been unlucky to get caught, as his intention was to capture Billy Woods in the *Northern Jewel*. As he could see the *Jewel* was getting away he grabbed the nearest ship – us. The skipper was brought back to the *Bombardier* with an officer from the gunboat who stayed on the bridge to make sure we didn't try to escape. We were escorted, under arrest, into Seydisfjord. The following day the skipper was taken to court and the ship fined for being inside the fishing limits.

It was 18 March 1958, my birthday. I was nineteen years old. We stayed in Seydisfjord till we were released and allowed to resume fishing. The local Icelandic newspaper reported the following (roughly translated):

Seydisfjord, 17th March (1958)

Aegir brought to harbour a big vessel, the Grimsby trawler *Bombardier* caught for illegally fishing within the fisheries limit in the early hours of Monday. The *Aegir* came upon the trawler about 1 nautical mile within the line in Lonsbugt, between Eystrahorn and Vestrahorn. The skipper of the trawler, a man by the name of Meech, admitted the violation and position determined by the coastguard vessel captain Thorarinn Bjornsson and his men. The *Bombardier* had about 600 kits of fish. A ruling in the case is expected tomorrow, Tuesday. The *Bombardier* is the first trawler to be caught for violating the fishing limits this year.

A couple of months later we were fishing on the east side of Iceland. The ground was very rough and we were sustaining a lot of damage to the net. The skipper was threatening to rig up the port side fishing gear which meant we'd be working eighteen hours a day every day.

Next day we were still in the same place. It was a lovely summer's day, the sea like a millpond. The fishing was good and consequently a lot of trawlers had congregated in the area, all outside the fishing limits for a change. The ships were working a small area and as usual getting in each other's way.

We'd only had our gear down for about fifteen minutes; we were all in the pounds gutting fish. I was gutting in the after pounds, I could see the Hull trawler *Othello* laid about half a mile away with her trawl doors hung up in the gallows ready for shooting her nets away. I thought she was waiting for us to get past.

I was surprised to see black smoke coming from her funnel. This meant the skipper had ordered the engines to full speed. I could see by the way she was turning she was going to hit us. I shouted to the lads to hang on. *Othello* hit us at speed, but because she was turning at the time only gave us a glancing blow, which was a good job because she had a very sharp bow. Still it inflicted serious damage. She hit us on the corner of the bow and as she slid along our port side she hit the fore gallows, smashing all the strengthening bars. Then she continued to slide down, pushing the ship's rails down as she went. She then hit the after gallows, smashing the trawl door in half.

When it happened there were three men working down in the fish room. I've never seen three men move so fast in my life. They came out of there like rabbits being chased by a ferret. The *Othello* did us a favour really: with so much damage to the port side it was impossible for the skipper to even contemplate putting the port side gear over.

Later the skipper of the *Othello* told our skipper that at the time he rang his ship on to full speed he was looking out his port side door, shouting a message across to one of the Hull trawlers who was close by him. He didn't realise we were there until he actually hit us.

I don't know what the outcome was, but if the skipper of the Hull trawler didn't have his master's ticket suspended then he'd at least have received a severe warning from the DTI.

CHAPTER 19

NORTHERN FOAM, 1959

My only trip in the *Northern Foam* was a nightmare, though as it turned out it became one of the easiest trips I ever did. We sailed for the Norway coast fishing grounds. The first day the skipper was down for his evening meal, but after that I only saw him about three times the whole trip. He spent all his time in his cabin on the bottle, all caused by his wife leaving him. He was given several warnings from the shore radio stations to stop using bad language and to stay off the air. He terrorised the radio operator to such a degree that he moved out of his cabin, which was next to the skipper's, and slept aft with the rest of the crew.

That trip the weather was appalling, it was blowing a gale of wind for most of the time with a couple of storm force 10s thrown in for good measure. On arriving at the fishing grounds the weather was too bad to fish so we lay to until it passed. I was on watch with the mate from 24.00 until 06.00 in the morning; the weather was so bad that each day we tried to scramble two hauls in. We'd call the crew out for breakfast at 06.00 then try and get two tows in while there was a bit of daylight in the sky. Bear in mind we were inside the Arctic Circle and only getting about four hours' daylight. Once breakfast was over it was my turn to go below. As I was rolling in, the rest of the crew was going out on deck to get the nets over the side

One of the few times I saw the skipper was when we went into Tromsø for repairs. He insisted he didn't need a pilot as he knew the way in, which he probably did when he was sober, but he wasn't. We headed into the fjords, the wind blowing, and there were heavy snow showers. We were called out to get the ropes ready for tying up and then sat in the mess deck for the next half hour waiting for the order to go on deck. While we sat talking, somebody remarked on the time it'd taken to get here, and the good speed we'd made. 'It usually takes longer than this,' one remarked. He turned out to be right.

The order came to standby the ropes. I went on the bow with the mate and two deckhands, the wind and snow made it very unpleasant and it was freezing hard, the visibility poor. We could just make out the lights on the shore when the snow eased off, then it would snow harder and we couldn't see them at all. We just peered into the darkness looking for the quayside. The snow suddenly eased off again and visibility improved dramatically. Ahead of us was a huge reef with large seas crashing over it, a frightening sight to which we were very close, in fact too close for comfort.

There were no signs of any town or harbour, just a few scattered lights of the houses on shore. The mate screamed out to go full astern; he shot off the bow and raced for the bridge. The engines were put in reverse. We were so close to the rocks we hung on to the handrail expecting to hit at any moment. What passed through my mind wasn't what I could see, which was terrifying enough, but what the hell was underneath us. I'd a terrible feeling we were going to run aground, and then we'd really be in serious trouble. The ship shuddered and shook as the engines started going full astern. It seemed ages before we were finally back into deep water and safety. It turned out we still had more than 10 miles to go before we reached Tromsø. The mate stayed on the bridge until we got there.

After safely tying up in Tromsø we discussed what we were going to do as we weren't very happy with the way things were. We informed the mate that if the skipper didn't sober up, then he could take the ship back to Grimsby. The mate eventually got him sobered up and the repairs were carried out. When we left Tromsø for the fishing grounds it was a relief to see a pilot on board to take us out!

Once back at the fishing grounds the skipper was back in his berth and back on the bottle so we were back to square one. We chatted in the mess deck as to whether we should carry on or get the mate to phone the office, explain the situation and take the ship back to Grimsby. It was decided we carry on as we all wanted to get a pay day out of it. The mate was an experienced skipper and well liked, we had faith in him to get a trip out of this mess. The weather was just as bad so we carried on where we'd left off, trying to get two hauls in each day.

We were towing the nets on the seabed for three hours each tow. On more than one occasion, by the time the net was hauled in the weather was so bad it'd have to be taken back on board and lashed up. By the time I turned out the fish was already in the fish room and everyone was off the deck. On other occasions we managed to get two hauls in so I'd have to go on deck to haul in. I hardly did any work at all that trip, but we didn't catch much and the proceeds were poor.

The skipper managed to surface for the last twenty-four hours on the fishing grounds. When we started steaming for home he instructed the mate that when we passed over the Viking Bank, if there were any German trawlers fishing there he wanted to find out what they were catching. If they were getting any fish we would have another twenty-four hours' fishing.

We passed over the Viking Bank about 03.00 and it was teeming with German trawlers. We went alongside one that was hauling his nets. We could see his net was full of red fish. The mate contacted the German trawler and asked if there were any cod or haddock amongst the fish; the German's reply was negative, only red fish. The mate threatened us not to tell anybody we'd seen these ships. We carried on steaming for home, hoping the skipper wouldn't turn out and see them.

Red fish didn't make very good money so from the crew's point of view it would've been a waste of time stopping, but from the skipper's point of view it would mean another 200 or 300 kits of red fish on his fish tally, which would make the trip look a lot better than it was. But did the skipper really deserve it?

CHAPTER 20

NORTHERN SCEPTRE, 1959/60

I joined the *Northern Sceptre* in March 1960. She was an excellent sea ship and comfortable. Vic Meech was skipper and it was the fifth time I'd sailed with him.

The *Northern Sceptre* was sister ship to the *Northern Jewel*. Both were identical but I think the *Northern Jewel* had the edge. Or maybe it was just that the *Northern Jewel* was the biggest ship I'd been in at the time. It was the top ship, the one everyone wanted to be in, what with all the excitement of poaching plus the large amounts of fish we caught. I can still see that trip in my mind when we caught all those large hauls of cod. It's something that I'll never forget. I've never seen hauls of fish like it before and I never saw them again.

The Norwegian coast was still in season and there was still plenty of fish being caught. The weather was fresh with a moderate swell and the day was clear, we could see for miles. The sun was very low and bright. We'd just hauled a good bag of fish and the skipper decided we'd do a short tow before steaming back.

It was a pleasant day and we were all in a good spirits, chatting about the good fishing and laughing and joking about things in general. While gutting the fish, we could see several trawlers steaming past us. There was a big modern Hull trawler astern of us, the *Cape Columbia*, steaming at full speed, I estimated about 13 knots as we were towing our gear and only doing about 4.5 knots. At first we didn't take too much notice as there was nothing to worry about. As she got nearer she didn't seem to be doing much to alter course and avoid us.

We shouted to the skipper to make sure he was aware of the ship behind us. He assured us he'd seen her and everything was OK. So we carried on gutting. The *Cape Columbia* was still coming at us. By now we were getting a bit worried. We could see the ship's wheelhouse and the two men on watch stood looking out of the window. We thought they must surely know we're here, but they kept coming. Now we'd stopped gutting and watched as the ship rapidly approached. We again shouted to the skipper; when he realised how close she was he blew the ship's whistle, but it was too late, the Hull ship hit us on our stern with a mighty bang; we could hear the sound of steel plates being torn apart as the ship heeled over to starboard.

We dashed aft to see what damage had been done. It was pretty bad. The entire after end had been split open, all the hand basins in the drying room had been smashed down to the deck and the accommodation on the port side had been partly demolished. Fortunately there was no damage below the waterline. After assessing the damage the skipper decided that due to the weather being in our favour we could continue fishing for another two days before going into Tromsø to get the stern patched up in case we encountered bad weather on the journey home.

The skipper of the *Cape Columbia* told Vic there were two men on the bridge at the time of the collision and neither had seen us until they actually hit us. They blamed the sun as it was low on the horizon and very bright. They suffered no serious damage so didn't need to go in for repairs and carried on with their trip. There would have been a Board of Trade inquiry but I never did hear the outcome.

We covered the damage as best we could with tarpaulins and cowhides. On arrival at Tromsø the shore gang came aboard and started cutting away the damaged plates around our stern. We had nothing to do as we'd stowed the nets and cleared the decks on the way in. All that was necessary to do was clean the accommodation for the journey home.

Some of the crew went ashore, bought bottles of brandy and sat in the after cabin drinking. I wasn't a big drinker; apart from my dram of rum and a couple of cans of beer dished out with the bond, that was it. I was offered a can of beer and a dram of brandy. I sat with the crew drinking it. I don't remember being drunk or even feeling merry. I remember everything up to about four o'clock in the afternoon and then it's a complete blank. It was like somebody had switched me off. I awoke at 03.00 the following morning.

I couldn't move. It was dark and absolutely freezing. I could feel a lot of weight on top of me. I was cold and scared. I had no idea where I was and started struggling to get free. Once I'd managed to climb out from under all the weight I realised I'd been in one of the bunks in the cabin that had been partly demolished. The crew had thrown mattresses, pillows, boots and anything else they could find on top of me.

I sat in the cabin collecting my thoughts. I couldn't remember anything since yesterday afternoon. After I'd pulled myself together I went to the galley. I felt awful, my head was throbbing. One or two of the crew were there and asked how I felt. 'Bloody lousy,' I told them. They laughed but I couldn't see the funny side of it at all, especially when they started telling me what I'd been up to! I thought they were pulling my leg, but apparently I'd ripped all the shelves out of the bunks and I threatened to fight with the biggest member of the crew. While I was in the cabin the galley boy came in to see if I was OK. I asked him if he'd seen my white thigh boots and when he said he couldn't find them I told him he wasn't getting out of the cabin until he did find them. He soon found a pair of boots and then dashed out of the cabin. All this was completely out of character for me. I'm normally an easygoing person who wouldn't harm a fly. I later found out the brandy we were drinking wasn't quality brandy but some cheap firewater they sell in Norway. To this day I've never drunk brandy again!

After the work was finished we continued our journey through the fjords. We dropped off our pilots at Lodigen and picked up two new ones to take us through the south fjords. The scenery is some of the best in the world. People, who go on cruises in summertime to see the midnight sun miss out in my opinion. The best time is winter when it's all covered in ice and snow, the houses you see scattered along the fjords are brightly coloured and stand out against the white, and to see a waterfall cascading down the mountain frozen in time is pure magic. Another great sight of the Arctic is the Northern Lights. To see these on a clear night is marvellous, the colours that flash across sky are truly unbelievable.

We dropped our pilots at Ålesund and steamed across the North Sea to the Humber Light Ship at the entrance to the River Humber. The passage across the North Sea was uneventful and the weather was good all the way home.

During the summer the skipper had a couple of trips off. The new skipper was from Fleetwood. On the way down to the fishing grounds the third hand was hitting the bottle which didn't go down very well with the new skipper but by the time we reached the fishing grounds he'd sobered up and was back to normal. Halfway through the trip he hit the bottle again. Twice he failed to tie the cod end up properly and each time they came adrift and the net was empty. On the third haul, he was going to tie the cod ends up and put them over the side when the skipper called out from the wheelhouse, telling him there was a hole in them. They had a big row, the third hand insisting in a slurred voice that there were no holes in his cod ends. 'I never get holes in my cod ends,' he said. The skipper made him hang the cod ends up so he could go inside and have a good look around for holes. We were all in stitches it was so funny. The last straw was when the third hand shouted from inside the cod ends, 'I told you there were no holes in my cod ends skipper,' and with that he fell out of a hole that was big enough to get a double-decker bus through. We all fell about the deck laughing but unfortunately for the third hand the skipper couldn't see the funny side of it and sacked him. It was a shame really because he was a good third hand and getting drunk was something he didn't normally do.

For the rest of the trip the crew was egging me on to ask the skipper if I could have the third hand's job. I wasn't sure about it. I could do the job OK but it seemed a daunting task. Then the mate told me to go ask the skipper for the job. 'Well if it's OK with you I'll go ask him,' I said. When I asked the skipper he didn't hesitate. He said yes.

The following trip was a bit of a nightmare. The first twenty-four hours were easygoing. We caught plenty of fish and the weather was fine. But like all good things it didn't last. The fish took off so we changed fishing grounds and steamed to Grimsey Island on the north coast of Iceland. The fishing wasn't good there either. We were catching lots of small red fish which got stuck in the meshes of the cod ends. The skipper wouldn't let us pull them out so the cod ends got clogged up with them. They got so heavy it became difficult for me to tie them up properly and I had to heave them out on the Gilson. This took too long. On returning to Grimsby the skipper told Vic that although I knew the job quite well I wasn't big or strong enough to do it. A fair comment really, I was still only twenty and wasn't very big at the time. I still think if the skipper had allowed me to pull those red fish out the cod ends each haul that I wouldn't have had any problems, I could have managed the job. But I'm still proud to say I was third hand of one of Grimsby's top trawlers at the age of nineteen. When we docked, Tommy Rowson offered me a job as third hand on the *Northern Pride*. 'It'll give you a bit of experience,' he encouraged. I had to think about it but I felt confident enough and agreed.

CHAPTER 21

NORTHERN PRIDE, 1960

And so I joined the *Northern Pride* as third hand in January 1960 with skipper Eddie Hall and mate George Turrell. George and I had sailed together on ships including the *Northern Sky*, *Bombardier* and *Northern Sceptre*.

I only did three trips in the *Pride*. She was a happy ship, everyone got on well, but after three trips I left. It wasn't that I didn't like the ship or any of the crew, but I had so many scrapes with death on her it was unbelievable!

One trip we were lifting a large bag of fish on board. As the bag swung inboard the Gilson wire parted and about 2 tons of fish crashed onto the deck, missing me by inches. Another incident was when we had what we called a one-ended job where the towing wires had parted. To get the gear back we had to heave all the gear up on the after derrick. Being an old ship she still had the original wooden derricks, which were huge, very thick, and weighed a ton. We were heaving the gear up when there was an almighty crash. The stay holding the derrick up had parted and it crashed down onto the ship's rail, again just missing me. It was so close that the man working the winch thought it had actually hit me. Another 3in closer and it would have killed me stone dead.

Another incident was similar to the last one. We'd another one-ended job but this time all the wires on the after end had parted. The weather was bad and we were pitching and rolling about. We were using the tackle to heave the gear up. We'd hooked it in the bobbins and were in the process of heaving them up. There were 30ft of these bobbins and there's an awful lot of weight in them. Every time the ship rolled to port the strain on the tackle is tremendous. The health and safety people running around these days would be horrified at the weights we used to haul up the mast. We had the bobbins hove right up to the top of the mast. I was stood underneath trying to lash them down to the ship's rail with chains. The ship lurched violently to port. I sensed something bad was about to happen and I could hear the wires singing as I jumped over the after deck boards to get out of the way. The strain on the wires was so great they all parted and 30ft of iron bobbins plus the tackle block came crashing down, hitting the deck where, seconds earlier, I'd been standing.

One of the deckhands shouted, 'Bloody hell Jim I'd get out of this ship before she gets you!' I took no notice. You put it down to being a part of the job. I've seen it happen many times.

It was the third trip that made me think perhaps he was right. The weather had been bad with plenty of wind and icing up. That morning the weather was about force 7/8 with a heavy sea running and a moderate frost. We had a thin coating of ice on the ship but not enough to get concerned about. We hauled the nets about 08.30 and had them back on the seabed by 09.00. From then on the weather rapidly deteriorated. At 10.30 the weather was so bad the skipper ordered all hands on deck to get the gear back on board as quickly as possible.

The weather had reached storm force 10 with wind speeds up to 60mph, driving snow and seas of up to 30ft. The seas were steep, with heavy spray being blown off their breaking crests. Visibility was reduced to about a mile. The ship was rolling and tumbling about and being pounded by huge seas. We scrambled from our bunks and got our sea gear on.

When I got on deck I couldn't believe the weather had become so bad in such a short space of time. As I stepped out the full force of the wind hit me and it was so cold it took my breath away. I was bent over double and leaning into the wind just to stay on my feet. The spray was hitting my face like frozen needles and I had to squint to be able to see anything. It was just horrendous.

The ship was laid broadside being pushed along by the wind and high seas. As the after door came to the surface it was being pulled away from the ship's side and the net could be seen lying on top of the water. The pull on the gear was immense and it took us a good 10 minutes to get the after door up into the gallows and unclipped.

This is a dangerous job in fine weather let alone when it's blowing. We'd safety bars to stop the door swinging inboard, but they were about as useful as a snowman at the equator! In bad weather it would swing round and smash the safety bars or unship them and they'd drop into the sea. You had to be careful not to get caught when it was crashing about as if you did chances were you'd lose an arm or get crushed and killed. Over the years many men were seriously injured or lost their lives working on the doors.

We really struggled to get the first part of the gear on board and lashed down so we could start pulling the net in. This is hard as we had to pull part of it up by hand. The crew stood along the rail grabbing the net as the ship rolled to starboard and when the net went slack we pulled like hell to get as much in as possible. As she rolled back to port we jumped on it, hanging on, to stop it being pulled back over the side. We had to take care not to get our fingers caught in the mesh or we'd go over with it. Sometimes the pull of the sea was so great we'd let it go and start again. After three or four attempts we managed to get enough net aboard to get a rope made fast around it. Once done the job became a bit easier. By heaving on the rope with the winch we managed to get half of it on board.

Suddenly from the bridge window the skipper shouted, 'Look out there's a sea coming!'

I looked up. 'F★★★ing Jesus!' I thought.

A massive sea was bearing down on us at great speed, at least 20ft high with a great white mass of breaking water on top of it. Everybody made a run for it, trying to get out of the way. It crashed down onto the deck, scattering us all about. I was up to my waist in water, clinging onto the winch.

Two crew were knocked over and washed along the foredeck. Fortunately there were no injuries but we were all wet through. You can't go off the deck to get changed in a situation like this. We were fighting for the safety of the ship and our lives. The sooner we got the net on board and lashed down the sooner we could get the ship's head up into the wind and sea and be reasonably safe. Chopping the gear away was never an option.

As the water cleared the decks we jumped back down and started from where we left off. We got another rope around the net and had just started to heave when again from the bridge came the dreaded cry, 'Look out, water!'

This time I didn't look up, I didn't want to see. Everybody scattered. As I jumped back from the ship's rail I felt the net running up the back of my leg as it streamed over the side. I went with it into the ice cold raging sea. The man on the rope hadn't made it fast in his haste to get out of the way. As the net streamed back over the side the rope flipped off the winch drum and-streamed back into the sea taking me with it! It happened in seconds.

'F★★★ing hell, this is it!' I thought. I was desperately trying to hold my breath. My eyes were open but all I could see was masses of bubbles. The noise was incredible. I could hear water rushing around and things banging against the ship's side. It didn't seem real; it was my worst nightmare. This couldn't be happening to me!

Had I tried to swim or struggled my gear would've filled up with water and I would've sunk. I didn't panic but just lay in the water not moving a muscle. All the time I was being pounded violently against the ship's side. I thought I was going down underneath the ship and would come up the other side. This worried me, but that was stupid, I'd be dead long before I surfaced on the other side.

I held my breath for as long as possible but then I started gasping for air. All I got was mouthfuls of ice cold water. It was going down in quantities, there was nothing I could do, I was no longer in control.

'Bloody hell,' I thought. I didn't want to die, I was too young. My thoughts were for my wife. We'd only been married about twelve months and it looked like she was going to be a widow.

My lungs were filling up with water; this was definitely the end and I was helpless to stop it. It felt as if a steel band was tightening around my chest, squeezing the life out of me. It was a strange feeling, uncomfortable but not painful. Then the noises started to drift away into the background. The darkness descended and I lost consciousness. I thought I was dead.

Seconds later I broke the surface. A 60mph freezing cold wind and icy spray blasting my face snapped me back to reality; suddenly I was alive again! I was back from the dead. I was choking and frantically gasping for air. I opened my eyes and all I could see was net, it was like being in a darkroom with luminous netting hanging up in front of me.

I wasn't fully conscious but I knew I had to get hold of that net or I was well and truly lost. I've never grabbed anything so hard in my life. I nearly broke my fingers as it was laid tight along the ship's side. I desperately hung on, not daring to let go. If I did I wouldn't have the strength to get back. I looked up to see the ship's bridge towering above me in the darkness, all the deck lights ablaze. It's not often you see the bridge from sea level – it looked massive – I could see heavy spray lashing against it. I closed my eyes and prayed it'd still be there when I opened them again. I was exhausted. I wondered how much longer I could hang on, though I knew it wouldn't be for long.

Suddenly I felt someone grab the neck of my oilskins. I opened my eyes to see a face peering over the rail at me. I can't describe how relieved I was. I couldn't speak, but inside I was begging him not to let go of me.

'He's here! I've got him!' I heard him shout.

Then I completely blacked out again, letting go of everything. I nearly pulled the lad over the side with me, but he managed to keep hold of me. If he'd let go that would have been the end. There would have been no getting back as I had nothing left to give. I was completely exhausted. Next thing there were hands grabbing at me from all directions. I just heard George shout, 'This time lads when I say pull!' I didn't hear anything else after that but I vaguely felt myself being hauled bodily onto the deck.

When I came round and managed to stand I struggled to stay on my feet and was clinging to the winch barrel. The weight of the water in my gear was incredible. If I'd let go I'd certainly have fallen over. I looked around at the crew. They were as white as ghosts. I think they were more scared than me! All were relieved to see me back on deck, but none more so than me. As I clung to the winch all I could say was, 'F***ing hell! F***ing hell!' I said it over and over, and each time the water was spewing from my mouth. I was virtually pumping myself out.

When I thought I was capable of standing on my own two feet I let go of the winch and staggered aft, falling twice on the way. I made my way to the galley to get warm. As I walked in the cook said to me, 'It looks like you got wet Jim.'

'Wet!' I said. 'I've just been washed over the f★★★ing side!'

'F★★k off Jim! You don't go over the side in this lot and get back!'

He was amazed when he realised it was true. He got me a pot of tea and a fag and I sat smoking and thinking about what I'd just gone through. I still couldn't quite believe it.

I went down to the cabin to get changed. I was shivering and as I took off my wet gear I could see the right-hand side of my body and leg was battered and bruised from being dashed against the side of the ship. I felt stiff and sore.

I dried myself off and put dry clothes on. I sat on the seat locker smoking another fag, gathering my thoughts together, when the cook shouted out down the ladder, 'The skipper wants you on the bridge Jim.'

I went to see him and he asked me how I was and gave me half a mug of rum. I felt a lot better after that and was beginning to feel a bit warmer. The skipper suggested that I turned in and got some rest. I didn't want to get in my bunk, I wouldn't have slept anyway. I went aft to the drying room, put some dry gear on and went back on the deck.

By now the gear was on board and lashed up and the ship's head was up into the wind. It was midday and the crew went aft for dinner. I worked on deck for half an hour or so just pottering about. I thought it was the best thing to do.

We left the Russian coast and steamed for the west coast of Norway.

Later that evening the mate told me more about what'd happened earlier. Apparently the sea that hit washed everyone round the deck but by the time the water had cleared and they were able to get to the ship's rail I was gone. I was nowhere to be seen. They scanned the sea looking for me. The first sighting was when the top of my sou'wester bobbed up to the surface, only to sink below the waves again. This was halfway between the bow and the bridge. Next time they saw me I was level with the ship's bow. George was convinced it was the last time anyone would see me.

Once you pass the ship's bow in weather like that you don't usually get back, you and the ship drift away from each other at a rapid speed. They were surprised when they heard the deckhand shouting he had hold of me. How the hell I got dragged back from the bow of the ship to the bridge amidships I'll never know. That must go down as one of the mysteries of the sea – but one I'll always be thankful for.

The deckhand who found me had been working aft when the sea hit us. When the water cleared he was making his way forward to help the crew search. He happened to look over the side as he was passing the winch and spotted me hanging onto the net. It all happened in about ten minutes; I was in the water for five. In Arctic conditions, four minutes is about maximum survival time. I was very, very lucky to be alive.

The following day we arrived on the west coast of Norway. The weather had calmed down and fishing resumed. I was feeling sore and stiff, but you just got on with it.

The rest of the trip George kept asking if I was OK. I told him I was, but he took some convincing. I wasn't having any problems, and I wasn't worrying about it. As far as I was concerned it was a bad experience that was now a thing of the past.

Two days later I was on the bridge taking the afternoon towing watch. At 14.00 hours the BBC shipping forecast came on. The forecast for our area was NW gale force 8 to severe gale force 9. I worried, but only for a couple of minutes and the fear passed. Thinking back on it, lady luck certainly smiled down on me that day.

By the way all that bullsh★★ about your life flashing before your eyes when you drown isn't true – well, I didn't experience it anyway – but I decided that the *Northern Pride* wasn't the ship for me after all so I left and joined the *Northern Sky*.

CHAPTER 22

NORTHERN SKY, 1960

I joined the *Northern Sky* at the beginning of September. The skipper hadn't been in that role very long. The last time I sailed with him he was a wireless operator. It was this trip we got caught in some of the most horrific weather I've ever worked in. We were fishing at the Barents Sea, along the Russian coast, and the weather had been bad most of the trip.

That morning we'd just shot our nets away, the weather really screaming, with huge seas and terrific winds. We didn't have much fish to gut and the weather was gradually getting worse. It only took us about three quarters of an hour to get the fish gutted and stowed away in the fish room, but a frightening forty-five minutes it was! We had the wind on the port bow and we were taking a real hammering. The seas were very high with huge crests breaking and they were running along like steam trains. It was difficult to stand up, the ship was struggling to make headway and the huge seas were crashing against the port bow sending heavy water cascading into the front of the bridge. It took us all our time to get back to the after accommodation.

We sat in the mess deck. We didn't think it worth going for a lie down as we expected the skipper would call us out on deck to get the gear back on board. We were wrong though. We towed the nets for a full four hours. When we did haul the nets, the weather was unbelievable, the seas were mountainous and the wind had now reached hurricane force.

The ship lay broadside to the weather and was being driven along sideways by the sheer force of the wind and waves. One minute we'd be perched on top of the sea looking at the horizon, the next we'd be at the bottom of the sea looking up at the next sea rolling towards us. It was like looking up at the side of a mountain, but these mountains were on the move, with huge seas breaking at the crest. When I got washed over the side of the *Northern Pride* the weather was nothing compared to this and I must admit I was scared.

We struggled to get the gear safely to the ship's side. We got fore and after doors safely secured in the gallows and started heaving the net in. As the end came into the gallows we watched as a huge bag of fish broke to the surface. We're going to have trouble getting this lot on board, I thought.

The force of the wind and seas driving us along was so great that as we got the net to the ship's side it started to rip away. It started at the after end and gradually worked its way to the fore end. It took about five minutes for the whole net to be ripped away. We could only stand and watch as net and bag of fish rapidly disappeared over the next sea never to be seen again. What a relief to see that net disappear though; it certainly made the job a lot easier, all we had to do now was heave the bobbins on board which took us about fifteen minutes. We lashed what was left of the gear down and made everything seaworthy.

The skipper sent us a dram of rum down from the bridge, it was the least he could do. The third hand was having Christmas at home and I was going to ask the skipper if I could have his job for the Christmas period. But after what we'd just come through I decided I too would be having Christmas at home!

When we docked I informed the office that I was going home and then I'd be going to the nautical school to sit for my mate's ticket. There I met the skipper who'd given me a start as third hand in the *Northern Sceptre*. I told him I was sitting for my mate's ticket. He asked how long I had to go, I told him three weeks. 'I want a new mate when we come in if you fancy coming with me?' Well there was only one answer to that!

I passed on Wednesday 13 March 1961 and went to see the skipper the day he docked to make sure he'd not had a change of mind. I asked him if Billy Woods would let me have the job as he was now working in the office and was in charge of signing on the mates. 'Don't worry about that,' he assured me. 'I'll go see the boss upstairs.' He came down and told me to go and sign on. Billy Woods was furious and made it plain that he was far from happy.

I sailed as mate on the Grimsby trawler *Serron* on Saturday 16 March.

CHAPTER 23

SERRON, 1961

My first trip on the *Serron* was the most daunting I'd ever done. When I arrived on board there was only the skipper I knew and looking around the crew there was only the decky learner who was younger than me. The crew was all very experienced, could probably tell me how to do the job and some of them tried. When we got out the dock we got the ship ready for the journey to Iceland. The skipper told me to take the first watch. The weather was fine, good considering that it was still only March.

After my watch I sat in my berth. It was the first time I'd had a berth to myself, it felt great and I sat there thinking about what was to be done on the deck the following morning when there was a knock on my berth door. I called out to come in. It was the fish room man. He told me he had been in this role for the last eighteen months and if it was it OK with me he'd like to carry on. I was delighted by this, as the fish room man is in charge of the fish room for twelve hours each day, when the mate isn't on the deck. The fish room was poor and not very good at keeping fish, which was bad news. On this ship there was a chilling plant for the fish room which was usually put on halfway through the trip.

The following morning we called all hands out for breakfast and started getting the gear ready for the trip ahead. I had no problems as I was pretty good at getting the gear together. The main thing was letting the crew know it, and getting them to do it. I think most of the crew thought that I was some young kid who didn't have a lot of experience. They were right, as I'd only ever done six trips as third hand. Most mates usually do a year before they are taken on as mate.

As the day wore on everything went well. I think I'd gained the confidence of most of the crew but there were a couple who resented someone so young telling them what to do. Apart from that I did OK.

During the trip we had a couple of foul gears. When the net gets caught on an obstruction on the seabed it pulls the warps together; this brings the trawl doors together and they spin, putting turns up the warps. They can be complicated and difficult to clear, other times they come out quite easy, but it's the mate's job to get them clear.

While trying to get them clear one of the deckhands started telling me how to do it. I listened to him and he was right, but I'd a different way which I also knew worked very well. I told him I'd try my way first and if it didn't work I'd try his way, which didn't please him. Fortunately my way worked to perfection. Afterwards I got him on his own and told him his idea was very good and I knew it would've worked but I did it my way as if I was going to get the sack for making a cock-up then I'd only myself to blame. We got on great after that.

I was on the bridge with the skipper one time. We were talking about the job and he was drinking a can of beer. 'My beer seems to be going down fast but I don't really think anyone is stealing it because it's in my bathroom, and nobody goes in there except me,' he remarked. He couldn't blame me as I wasn't taking any spirits or beer that trip.

While towing our gear along I went for a walk round the deck to check everything was OK. As I turned to the starboard side I found out where the skipper's beer was going. As I stepped out from behind the winch I came upon two deckhands. They had a broom handle with a noose on the end of it and were putting it through the open porthole into the skipper's bathroom and lassoing cans of beer. They'd three cans out already when I caught them. I told them I could tell the skipper, which meant they'd get the sack, or they could take what they had and not to do it again, their choice. They chose the latter and offered me a beer, which I declined.

The fishing was very poor but the weather was unusually warm for the time of year. After filling up a pound with fish we'd shovel about 2ft of ice between the fish and the deck head to keep it nice and cool, but with the warm weather the ice was melting away too fast. The fish room man suggested it was time to put the chiller on but the skipper refused, saying that we didn't need it as it was only used for steaming home. The fish room man and I told him if we didn't use it we'd have fish going off, but he still refused.

The day we landed our fish I went down to the fish market to look at the quality of the fish. I was very pleased to see it looked OK. That morning there was a lot of fish on the market. The market was buzzing with fish merchants going round each ship checking the quality and which trip of fish they were going to buy from.

Before the sales started each trip of fish was inspected for quality and freshness by a man from the Fisheries Ministry. He was known as the pepper pot man; how he got that name I don't know, but he was a mate's worst nightmare depending on what sort of mood he was in. He could make or break you and was someone not to get on the wrong side of.

That morning we put 1,500 kits of fish on the market. The fish was presented in 10-stone boxes and laid out on the fish market in blocks of 200 boxes. On arrival at our trip the pepper pot man walked up to the first block of 200 kits and looked into three boxes of fish. He put his arms out and condemned the whole lot. I was furious and was going to have it out with him, but the fish room man warned me off. 'Do that and he will condemn more fish.' So that was it, I had to accept the fact that 200 kits of good fish had been condemned. I was totally gutted.

After the pepper pot man had gone I overheard two of the fish merchants talking when one turned and said to the other, 'Fancy condemning good fish like that, I'd have bought it.' I approached the fish merchant and asked what his opinion was of the quality. He said it was OK, not the best in the world but nowhere near bad enough to be condemned. 'I'd certainly have bought it all,' he repeated.

I got the sack, thanks to the pepper pot man.

When I reported to the office later that morning I was called in to Woody's office. He made it plain he didn't like skippers going to the gaffer over his head and signing on mates as it was his job. 'Go back to sea as a deckhand and get more experience as third hand. Then come back in three years and I might consider giving you a mate's job.'

I was absolutely livid. I told him, 'This time next year I'll have my skipper's ticket and I'm taking my ticket out of this office and changing companies. The next time I come into this office will be for a skipper's' job.'

I asked him to give me my mate's ticket and I left Northern Trawlers and joined the Ross Group. It was 5 April 1961 and the skippers were in a dispute with the owners. After a meeting there was still no agreement and the skippers came out on strike.

On reflection walking out of that office was a bad thing. What I should've done was gone upstairs and had a word with the gaffer.

CHAPTER 24

THE ROSS GROUP, 1961/2, *REPERIO*

On 13 May 1961 the strike ended. I was told I'd to go third hand on the trawler *Reperio* for a few trips and then they'd give me a job as mate.

She really was something else. She went back to the Dark Ages, a coal-burning trawler with the bridge aft side of the funnel. Built in 1907, she was 91 net tons with 70bhp. She was fifty-four years old and she certainly looked it!

She had no power to speak of when towing our gear. Even the seagulls paddled past us, and if the engineer let the steam pressure drop in the boilers she'd virtually stop. The first time it happened I was on the bridge – I thought we'd got our nets fast on the seabed. I rang the telegraph to slow speed and called all the crew out. The skipper strolled onto the bridge, had a look and told the crew to go back to their bunks. He turned to me and said, 'We're not fast, that bastard of a second engineer has let the steam drop back.' He told me he'd sack him but there weren't any decent engineers who'd sail in this old bucket. 'Give it half an hour then she'll start moving again as the pressure in the boilers builds up, make sure the gear isn't foul then carry on towing for another hour and a half.' That's great, I thought. It's a good job the weather is fine or the nets would have been in a right mess.

How these ships passed their safety certificates I'll never know, they were just about falling apart. They were steel boats and the plates were getting pretty thin. I once threw an iron bar on to the engine room casing and it went through the plates like a spear, they were that thin! Another time the scupper door in the side of the ship was jammed, I gave it a few hefty kicks to loosen it and the whole lot fell out of the ship's side and dropped into the ocean, never to be seen again.

When we sailed the following trip they'd just welded another scupper door over the hole where the old one used to be. I only did two trips on her then went mate on the *Ross Lion*.

The *Ross Lion* was a diesel-powered trawler, nice and clean. She weighed 274 gross tons and had 670hp. The skipper was Stumpy Reynolds and I got on very well with him. Not only was he a good skipper but an excellent fisherman. He knew the North Sea cod grounds like the back of his hand.

During this time the Ross Group was building new vessels especially for the North Sea and middle water fishing grounds. They were known as the birdie boats as each was named after a bird: *Ross Tern*, *Ross Falcon*, *Ross Kestrel*, *Ross Hero*, *Ross Hawk*, *Ross Cormorant*, *Ross Curlew*, *Ross Eagle* and *Ross Kittiwake*, and they were ideal ships for the North Sea and middle water.

The *Ross Lion* was an excellent ship when steaming before a gale of wind, but in every other way she was a terrible sea ship. She took more water on her decks than the *Titanic*! And I was washed along her decks on more than one occasion. I got on so well with the skipper he said when we docked he was going to ask the office if he could take me as mate with him in the new ship the *Ross Falcon*, which would be ready for him to take on her maiden voyage by the time our trip was over.

I felt very privileged and although I'd not spent a lot of time as a mate he still thought I did my job well enough to go in the new ship with him.

Unfortunately it didn't turn out as I'd hoped. When we arrived back in Grimsby and landed our fish the skipper asked the gaffer and he refused, their excuse being they wanted an experienced mate to go with him. I was told I would stay mate on the *Ross Lion*.

So on the next trip we sailed with a new skipper, Jim Gladwell. He was about twenty-three years old and this was his first regular ship as skipper. He was very keen to make a success of his new command. Jim and I were around the same age and became good friends; we sailed together for about eight months from September to the following April.

It wasn't a very good winter. We had a lot of bad weather and being in a ship like the *Ross Lion* made it hard going. Jim worked in a lot of bad weather, but was very careful and left me to get on with the job of running the deck. He didn't interfere and the work on the deck ran smoothly. I wouldn't say Jim was a stormy sod or anything but I'm sure if he took his shirt off you'd see he had dorsal fins down the back of his spine.

We had some good times together on that ship and some frights. On one trip we were on the Danish coast and were changing fishing grounds during the night. We were running before a strong gale of wind when the weather rapidly increased to a severe storm 10, the seas incredibly high.

We needed to get the ship round head to wind. Jim told me to take the steering wheel and wait for his orders. Speed was reduced to slow and we waited till the ship was reasonably steady, then Jim ordered me to put the wheel hard over to starboard. With that he put the engines on to full speed.

The ship started to spin rapidly to starboard, picking up speed as she went. As we were coming broadside to the seas we started rolling about violently. You don't realise how high the seas are when you're before the wind. These were mountainous.

We were nearly up head to wind when a huge wave loomed up and broke over the starboard bow. We ducked down below the level of the bridge windows, expecting them to be smashed when the sea hit us.

The whole ship shuddered and shook as the force of the sea crashed onto the deck, stripping the lights off the handrails and the foremast. One hit the bridge with such force it put a large dent in the bridge front. Later when the weather had fined away we found the light on the deck, it looked like it'd been run over by a steam roller!

The following day we heard the Hull trawler the *Artic Viking* had sunk. She was on her way home from the Norwegian fishing grounds in a northwest storm. She'd just passed Flamborough Head and was about two hours away from home. During the night the ship had to be stopped to repair a burst pipe in the engine room. As she was stopped broadside to the seas she was hit by two huge waves, one straight after the other. The ship heeled over and capsized as the second sea hit. It took only two minutes for the ship to go down.

Fourteen of the crew managed to scramble onto an open life raft but five men weren't so lucky and went down with her. The men in the life raft floated in mountainous seas for over an hour, desperately sending up distress flares. They were eventually seen by the Polish lugger *Derkacz* but it took over four hours for her crew to rescue the men due to the treacherous conditions. Each man was rescued from the raft by means of a rope. After the rescue the skipper of the *Derkacz* had to keep his ship heading into the storm. It was nearly two days before the crew was back in Hull.

On our next trip we were back on the Danish coast. The weather was very bad again, and while hauling our nets we took a huge sea that washed us all along the deck. It was quite terrifying. Being knocked down by a sea is no joke; you don't know where you're going to finish up. You can feel the power in the sea as it propels you along the deck. You just hope and pray you don't get smashed into the deck stanchions or the winch, or any other hard obstacle along the way. Or worse still get washed over the side.

As we were getting ourselves sorted out, we were told we'd a casualty. One man had been smashed against the winch, hitting his face on a protruding object. He was bleeding from one eye and it was swollen up like a balloon. I told a couple of the crew to take him aft and make him as comfortable as possible. I went to the bridge to inform the skipper what had happened. He went aft to have a look at the injured man and on his return told me to get the gear on board as soon as possible. When we'd done we'd make our way to the nearest port, Esbjerg in Denmark, to get him medical treatment.

With the gear on board and stowed away the course was set for Esbjerg. The skipper informed the port authorities we were on our way and ordered a pilot to escort us into the harbour. On arrival the pilot boat came out to us, but the weather was too bad for him to put the pilot aboard. We had to follow him into the harbour. The channel going into Esbjerg is very narrow with a sharp bend. As we followed there was a sudden violent snowstorm – we couldn't see a thing and lost sight of the pilot boat. The snow, driven by gale-force winds, was so thick it was impossible for us to see anything on the radar. The skipper eased the ship down to slow speed as we'd no idea where we where or where the pilot boat was.

A few minutes later the pilot boat called us up on the VHF radio telling us to alter our course to starboard, as we were getting too close to the edge of the channel. The skipper gave the order to bring the ship around to starboard. As he did so I glanced at the depth finder and to my astonishment the water was shoaling rapidly. I opened my mouth to tell the skipper but before I could get any words out we hit the edge of the sandbank. The ship gave a sudden lurch to starboard and heeled over at an alarming angle and we had to grab hold to stop ourselves being thrown off balance. Once we hit the deep water again she levelled up but it was a scary moment.

We sighed with relief as the snow showers lifted and once again we could see the pilot boat and what was on the radar. We went down the rest of the channel and into Esbjerg with no more trouble.

On reaching the harbour we were instructed by the harbour master to moor up at the quayside, where we were met by immigration officers, customs and a doctor. They ordered the skipper to hoist the yellow flag to indicate we were in quarantine. Jim asked the reason and was told it was because of the crewman with polio. At that time there was a polio outbreak at home and somehow they'd got their messages wrong. After a lengthy discussion and an examination by the doctor they allowed us to take the flag down.

The doctor examined the injured man; he was to attend the hospital for an X-ray to rule out any serious damage to his eye.

The immigration officers wanted to see all our passports, but nobody had one. They decided we could go ashore on condition that we didn't get into any trouble. Any trouble, they warned us, and they'd come down on us like a ton of bricks. We had a good crew and there was no trouble at all.

I was very impressed with the town; it was very clean, there was no litter to be seen and the people were very friendly and made us feel welcome. They were a credit to Denmark, I have to say.

We were pleased to hear that the crewman's eye wasn't as bad as we first thought. There was no serious damage. The hospital gave him some drops to put in it and told him not to work on the deck for a couple of days. Jim bought a chart for the channel out of Esbjerg, and decided not to take a pilot. The following morning we left the harbour. There were still a few showers around but the wind had eased off slightly at least.

A Danish fishing boat left the harbour with us and told us to follow him out, which was kind of him and we left together.

Jim told me to take the wheel until we were clear of the channel; the visibility was good for about five minutes, then the heavens opened. We were back in another snowstorm. The Danish fishing boat disappeared in the snow and the radar was wiped out. We had to rely on the depth finder to keep us away from the sandbanks but we still managed to hit the bottom twice. We were glad to get clear of the channel. It was a bit nerve wracking but we made it.

I knew one of our crew was going into the office and informing them what was happening on board during each trip. This was a regular occurrence on trawlers and runners would have their moles so they could keep tabs on everyone and everything. We had an idea who ours was so we set a few traps. We told the crewmen one or two lies and sure enough they got back to the office.

A couple of trips later I was called into the office and the runner told me Jim and I were being split up. 'Why would you want to do that?' I asked. The runner told me he'd been hearing things. 'What things? Can't you be more specific?' I asked. 'I think it'd be in your best interests if we leave it at that. I'd quit while your ahead,' he said.

I was taken out of the *Ross Lion* and went mate on the newly built *Ross Cormorant*. I'm sure if I'd not got the sack Jim and I would have spent a few more years together.

The *Ross Cormorant* was a modern trawler, only a few months old, and the skipper was Stan Shreeves. He was a good North Sea skipper and experienced on the Faroe Islands and the grounds north of Scotland. We got on OK, I suppose.

I wasn't very familiar with the skippers on these vessels, and when the crews spoke of different skippers I had no idea who was who. One afternoon I was in charge on the bridge while the skipper was having his afternoon nap when a conversation started on the ship's radio amongst the other skippers. They were calling some skipper called Rachmaninov, and boy were they calling him. Poor bastard, I thought, he must be the most hated skipper in the North Sea. An hour later when I called our skipper out to haul the nets up, I told him the fishing reports as to what other ships had been catching and just as I was about to leave the bridge I asked him who Rachmaninov was. 'I haven't a clue,' he said. 'Why, what have they been saying?' I told him everything and when I'd finished he said, 'OK you can bugger off now.' He seemed a bit grumpy, but

TRAWLER'S OTTER-DOOR JURY RUDDER.
Photograph taken on arrival home after a 300-mile passage
from St. Kilda.

1 A jury rudder rigged up
by my father on *Hayburn
Wyke*. (Author's collection)

2 *Comitatus*. (Peter Horsley
collection and Lancashire
County Council)

3   A large haul of fish – the sort of haul my father would have gone out for. (Photo Peter Brown)

4   An unwanted fish (basking shark). (Photo Peter Brown)

5   The ill-fated trawler *Goth*. (Peter Horsley collection and Lancashire County Council)

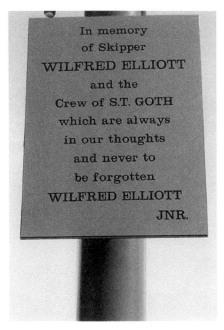

6 A plaque to *Goth*'s crew. (Author's collection)

7 *Goth*'s funnel. (Author's collection)

8 *St Leonard.* (Peter Horsley collection and Lancashire County Council)

9   Aerial photo of Grimsby Dock. (*Grimsby Telegraph*)

10   The dock tower. (*Evening Telegraph*)

11   Freeman Street party, with the dock tower in the background. (*Grimsby Telegraph*)

12   A busy day working on the North Wall, with Grimsby's top trawler, *Ross Revenge*, in the background. (*Grimsby Telegraph*)

13   The Grimsby trawler *Donalda* taking on coal at the coal hoist. The dock tower is in the background. (*Grimsby Telegraph*)

14   Ships on the slipway having repairs. (*Grimsby Telegraph*)

15 A ship on a slipway. (*Grimsby Telegraph*)

16 The notorious Lincoln Arms. (*Grimsby Telegraph*)

17 *Northern Chief*: looking down from the after-mast. (Author's collection)

18 Party on the fore hatch, *Northern Chief*, 1954. (Author's collection)

19 The author, far right, aged seventeen. (Author's collection)

20 *Northern Jewel.* (Painting by Steve Farrow)

21 The *Northern Jewel*. (*Grimsby Telegraph*)

22 The *Northern Jewel*. (Photo Fred Goodman from Janeen Willis collection)

23   The author, very tired, aged seventeen.

24   The author mending the net on the *Northern Jewel*, 1955.

25   The author taking a break.

(All author's collection)

26 Rum time! (Author's collection)

27 A good haul of fish on the *Northern Jewel*. (Photo Fred Goodman from Janeen Willis collection)

28 Another large haul of fish. (Photo Fred Goodman from Janeen Willis collection)

29 Severe icing, Iceland. (Photo Memory Lane)

30   Another bleak day in Iceland. (Photo Fred Goodman from Janeen Willis collection)

31   *Northern Sky*. (Painting by Steve Farrow)

32 Royal Naval Reserves on the *Northern Sky*, Chatham, 1956. The author is on the far left. (Author's collection)

33 *Northern Duke*. (*Grimsby Telegraph*)

34 The *Pelagia*, lost off Norway in 1956, seen here as the *Philip F. Thomas*. (US Coastguard and the Steamship Historical Society of America)

35 *Royal Lincs*. (Painting by Steve Farrow)

36 Dropping the bobbins in board. (Photo Fred Goodman from Janeen Willis collection)

37 Small icebergs, Bear Island. (Len Taylor)

38 & 39 Two photos of severe icing taken by Len (Chick) Taylor when he was engineer of the trawler *Black Watch* off the coast of Greenland in February 1961. (Photos Len Taylor)

40, 41, 42 Severe icing, Iceland. (Photos Memory Lane)

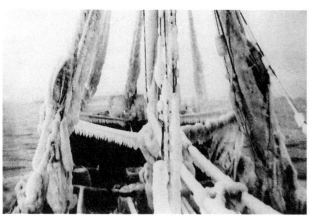

43 *Bombardier*. (Painting by Steve Farrow)

44 The author's wedding day, 21 March 1959. (Author's collection)

45 *Northern Sceptre*. (Painting by Steve Farrow)

46 Repairing the nets. (Photo Fred Goodman from Janeen Willis collection)

47 Overhauling the cod ends. (Photo Fred Goodman from Janeen Willis collection)

48 Shooting away nets. (Photo Fred Goodman from Janeen Willis collection)

49 Working on the fishing ground. (Photo Fred Goodman from Janeen Willis collection)

50 Working the winch whilst hauling the gear up. (Photo Fred Goodman from Janeen Willis collection)

51 One-ended job, after parting one of the towing wires. (Photo Fred Goodman from Janeen Willis collection)

52   Rough seas: a force 8 gale. (Photo Fred Goodman from Janeen Willis collection)

53   Pulling the net in. (Photo Fred Goodman from Janeen Willis collection)

54   The fish merchants jostling with each other to buy the fish. (*Grimsby Telegraph*)

55   *St Hubert*. (Author's collection)

56 Mine's a good 'un! (Bill Greene)

57 The trawler *Syerston*. (Author's collection)

58 A good haul of fish on the *Syerston*. (Author's collection)

59 And another good haul. (Author's collection)

60　A haul of large cod. (Author's collection)

61　The author in the foreground mending nets. (Author's collection)

62 The author talking on the radio. (Author's collection)

63 Pat in the galley. (Author's collection)

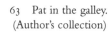

64 Pat steering the ship. (Author's collection)

65 *St Romanus*, lost with all hands – twenty men – on 11 January 1968. (Hull Marine Museum)

66 *Ross Cleveland*, lost on 4 February 1968 with all hands except the mate, who was found three days later cowering behind a building on the shoreline. (Hull Marine Museum)

67 *Kingston Peridot*. (Hull Marine Museum)

68   Severe icing, Iceland. (Photo Memory Lane)

69   Working the winch in icy conditions. (Susan Capes, Hull Museum & Art Gallery)

70　*Ross Tiger*. (Painting by Steve Farrow)

71　The *Ross Tiger* entering the fish docks, 1958. (Richard Holden)

72 *Osako.*
(Painting by
Steve Farrow)

73 *Ross Jaguar.*
(Painting by
Steve Farrow)

74 *Ross Jackal.*
(Painting by
Steve Farrow)

then he wasn't the most humorous person when he'd just been woken up. I left the bridge and went down to the mess deck. I told the lads about the conversation I'd had with the skipper and they all fell about laughing.

'Come on,' I said. 'What's the joke?'

'Don't you know who Rachmaninov is?' One asked.

'I've no idea,' I replied.

'Well you've just been talking to him on the bridge.' And he fell about laughing again.

Great, I thought, another good start with a new skipper.

The first trip we worked on the Danish coast fishing for sole at the Sylt Rough. We did well and returned home with a good trip consisting of 430 kits of plaice and seventy kits of Dover sole.

That morning I was down the market to see the fish landed and to give the shore workers a list of jobs that required doing. When I'd finished I went on the market to see how the landing was going on. There were only a few more boxes of fish left to come out of the fish room so I started to count how many boxes we'd turned out.

I started on the Dover sole; we had eighty-six boxes, sixteen over my original count so I was pleased about that. We also turned out more plaice. The total amount was 455 kits of fish all in excellent condition. Later that morning I went down to the office to get my pay. When you're mate or skipper you get a landing list stating how many kits of fish you turned out, and what price you've got for each type of fish. We made a good trip and everyone was happy.

Next trip I was sat in my berth studying the sale note when I noticed on the list we'd only landed seventy-five kits of sole. I knew this was wrong as I'd counted eighty-six kits myself, making us eight boxes short. I pointed this out to the skipper who told me to let the office know when we landed. The office said they'd look into it and the following trip on our sales note was the extra eight kits of sole from the last trip. If I'd not spoken up I don't think we'd ever have been paid for them.

I'm not saying there was anything dodgy going on, but you had to keep an eye on your fish; there were one or two unscrupulous firms who were known for ripping you off when you landed your fish.

It did seem a bit odd, however, that the following trip I got the sack. I was told the *Ross Cormorant* was going to the Westward, the name given to the grounds north of Scotland. As it was going fishing for dogs the firm wanted an experienced mate as it was a specialist job. Specialist job my eye, there was nothing special about catching dogfish. So I was back on the dole.

While I'm on the subject of dodgy firms I'll tell you about another. I was mate in one particular firm and had a friend who worked as a lumper: the lumpers land the fish. Trip after trip he kept telling me about the amount of fish the firm was taking off each trip; I used to tell him I didn't want to know as there was nothing I could do about it. If I'd started making a fuss I'd have only have got myself the sack again. As I was on a good ship with a good skipper and we were earning plenty of money I wasn't about to do that.

However, one trip on landing day I got up early just to see what was really happening. I arrived on the market at 05.30, about an hour and a half earlier than usual. At the ship was a lorry backed up to the market and two of the firm's employees were stacking boxes of fish onto it. When I asked what was going on they told me the gaffer wanted some for his fish house. I said I hoped it would be going on the landing list and they assured me it was. Fifty-eight 10-stone boxes of fish went off our ship to the fish house, all best quality too.

My next job was to count the boxes of fish we had on the market. There were 960 boxes and I checked it thoroughly as to make no mistake. When I got my landing list that's exactly what we landed, 960, no mention of the fifty-eight that went on the lorry, which would have made about £1,500. As the skipper was paid £52 to the thousand and I was paid £47 added to the money the rest of the crew lost they'd be earning hundreds of thousands of pounds tax free over the years. They had about twelve ships and it was happening on all of them. This was known as the ghost train.

I wasn't very happy with the way the Ross Group were treating me and was getting itchy feet. To sit for my skipper's ticket I had to complete twelve months actual time as a mate on board a trawler, the time in dock between ships didn't count. Up to now I'd done ten months and I'd four months left to get the other two in. It was going to be touch and go. I couldn't be messing about going third hand or having too much time ashore.

CHAPTER 25

SIR THOMAS ROBINSON & SONS, 1961

I met my friend Alfie Hodson who told me to go round to Sir Thomas Robinson's, Toshes for short, and ask them for a mate's job. He reckoned they were one of the best firms in Grimsby. I went to the office the following morning and met Harry Bradley, the ships' husband, and explained I was looking for a mate's job and was looking to get enough time in as mate so I could go for my skipper's ticket that coming September.

There were no mate's jobs available, but he had a third hand's job in the *Olivean*. I declined the offer, explaining that I was willing to wait ashore for a mate's job until I'd no time left to go and get my ticket. 'That's OK,' he said. 'Give me a look in the morning.'

Next morning I went to the office only to be told the same. Harry mentioned the job in the *Olivean* again, but I declined.

The following morning he had a proposition for me. 'Go in the *Olivean* for two trips third hand. The mate on the *Olivean* at the moment is going mate on the *Thessalonian* in a couple of trips. And if it's OK with the skipper you can go mate when he leaves. I'll make sure you get your time in to go for your ticket in September.' How could I refuse an offer like that?

I did two trips on the *Olivean*. Although I got on OK with the skipper, I wasn't too keen on him and didn't really rate him much, but as I'd be getting my time in to go for my skipper's ticket, it'd do.

We landed on that second trip with the *Thessalonian* and I was looking forward to going mate. When I got in the office Harry called me in for a word. 'Sorry I have to tell you this, Jim, but there's no mate's job. The gaffer wants the mate on the *Olivean* to stay where he is.'

I was bitterly disappointed but it wasn't Harry's fault. Harry promised me the next mate's job that turned up would be mine, regardless of what ship it was. He couldn't have been fairer than that. So I just had to be patient.

There were no ships landing the next day so I didn't go down to the office. On Saturday morning there was a knock on the door. It was Harry. 'I've got a mate's job for you,' he said. 'You're going mate on the *Thessalonian*, sailing Monday morning.' 'Great,' I replied and thanked him. It couldn't have been better. I told Alfie what ship I was going on. 'Great news Jim, but I don't think you'll get on with the skipper. He's a bit of a miserable old sod.'

The *Thessalonian* was a new diesel engine ship, very modern, less than a year old. When I arrived on board I put my gear in my berth and went on the bridge to introduce myself to the skipper but he was busy talking to the watchman, so I stood waiting until he'd finished. As the watchman walked off the bridge the skipper turned to me and said, 'Now then sonny are you one of the new deckhands?' 'No I'm your new mate,' I replied. 'Bloody hell you're a bit young aren't you?' Great, I thought, getting off to another good start!

The skipper, Lenny Coultas, was about fifty and didn't speak much to anyone during the trip. I'd go on the bridge to relieve him and he'd not utter so much as one word, he'd just stand at the wheel, sometimes for as long as an hour. Then he'd turn, say, 'There you are, the orders are on that piece of paper,' and go straight to his berth, and that was that. When he rolled out of his berth to haul the nets he'd simply say, 'Off you go then'.

It was a bit hard going to know what to say to him, so I just said nothing. Then one day out of the blue he started talking to me; I was really taken aback. He asked me all sorts of questions, including where I lived, and when I told him I lived about six doors away from him he was quite surprised. From then on we got on like a house on fire and became lifelong friends.

I'd been in the *Thessy* for about three trips when Harry Bradley called me into the office. 'The gaffer wants to see you in five minutes,' he told me. Great, I thought, now what've I done wrong? Five minutes turned into three quarters of an hour and by the time he called me in I was a bundle of nerves. He told me Harry had been giving him a good report about me. Then he spent the next half an hour or so asking me all sorts of questions about myself. Then he said, 'The reason I've asked you all these questions is because I've put you mate with Lenny Coultas, he's our best skipper and no one has the patience to work the grounds like he does. I want you to learn all his fishing grounds. I've had a word with him and he's pleased with your work as mate, and tells me you're eager to get on and become a skipper. He's promised to teach you all the grounds he fishes on. If you get your skipper's ticket in September you'll do a year with Len then the first regular skipper's job that comes along will be yours. Until then you'll be our relief skipper so you'll get plenty of experience.'

When I came out my head was spinning. Lenny Coultas was one of the top skippers in the firm. I just couldn't believe my luck. The downside was that ten years later I'd probably still be relief skipper as there were never any skippers sacked. I didn't mind really because I was doing as much time as skipper as I was as mate, and it was, as Alf said, the best firm on the dock. During the next ten years I sailed in all the ships in the firm except one, which was the *Gallilean* the other ships being *Olivean, Thessalonian, Judean, Phillidelphian, Tiberian, Pricillian, Rhodesian, Samarian* and the *Ephesian*. Some of the ships I sailed in more than once. I sailed on the *Thessy* six times during those ten years and at least twice on the others.

I won't tell you about the ships in order of the times I sailed on them or we'd have six chapters called *Thessalonian*. I'll do them in order, and tell you the experiences I had at different times in each one of them.

The first was the *Olivean*, not a ship I liked very much. In my opinion she was a bad sea ship and I was never happy in her. She was so bad she became known amongst the crews as 'the lump of iron'.

I was never at ease in the ship; I always had a feeling that something was wrong, or that something was about to happen. She was identical to the *Thessy*, but they were miles apart. The *Thessy* was a happy ship and I enjoyed every minute I sailed on her. When I was skipper on her we always did well. However, nothing much ever happened on the *Olivean* except one trip when I was skipper on her.

The worst part of being skipper on a trawler was the first two days after leaving dock, especially fishing in the North Sea. The main problem was booze. Fishing in the North Sea could mean putting your nets over the side a matter of hours after leaving the dock. If the crew was in a bad state I'd steam for three or four hours while they sobered up.

One trip we'd left the dock and got clear of the River Humber. The crew was in a bad way so I decided to do a long steam. I was on my own on the bridge deciding where to go when the second engineer came staggering onto the bridge. He was well pissed and as he came through the door he was followed by huge clouds of black smoke.

'Skipper did you know the galley is on fire?' he slurred.

'No! But I bloody well do now. Go down the engine room, stop the ship and get the fire pump switched on as fast as you can.'

I had a quick scan over the horizon to make sure there were no ships in the area in danger of colliding with us then left the bridge and went aft to the galley. As I got aft one or two of the crew where standing in the alleyway and I pushed passed them to get a look. My heart sank; the galley was one huge ball of flames. The heat was unbearable and it scared the hell out of me. 'How the hell are we ever going to get this out?' I yelled.

I ordered the crew to get the fire extinguishers; we were certainly going to need them to put this lot out, and cut the oil fuel to the galley stove. Then we attacked the flames. It was all hell let loose for the next fifteen minutes.

We lathered the galley with foam and eventually got the fire out, but it was more by luck than judgment. None of the crew had any firefighting training, including myself. It was frightening I can tell you!

The damage to the galley was superficial, mostly smoke damage, so we carried on to the fishing grounds and continued the trip. The crew had a bit of a moan about having to scrub the galley out, but apart from that it was OK. The cause of the fire was one of the crew had turned up the galley stove to make tea, then gone and left it. Consequently the oil had overflowed out of the drip tray and flooded behind the stove and burst into flames. What made it worse was as the wind ran along the alleyway into the galley it went underneath the stove, blowing all the flames up the back of the stove, curling the flames into a huge ball of fire round the galley.

It was years later when I went on a marine firefighting course that I realised a fire in a galley wasn't a serious fire at all. If we'd known how to put it out properly it would've been put out in minutes without all the panic. But you live and learn.

Another time we'd just left the river when the second engineer went to light the oil burner for the central heating. He was new to the ship and no one had told him when the boiler was lit there was always the danger of a flashback. To light the boiler he used a long taper with an oily rag. He had to put this taper though a small hole to light the oil jet under the boiler. And true to form he got a blowback – the flame shot out and ran right up his arm. He came onto the bridge and asked me to look at his arm, his shirt still smouldering.

When he took his shirt off I saw the state of his arm and told him I was taking him back to port for medical attention. He begged and pleaded with me not to as he'd been out of work for six months and was desperately trying to get some money so his wife and children could have a good Christmas. We only had two trips to do before Christmas. I thought about it and decided to carry on, on condition that if things went wrong he wouldn't blame me. He agreed.

In the medical chest we carried large gauze bandages especially for burns. I bandaged his arm best I could. The instructions were they'd to be left on for thirty-six hours. After the time was up I took them off and wasn't impressed with what I saw. His arm was in a terrible state, and I told him he really should have it seen by a doctor, but he insisted we gave it another twenty-four hours, so again I warned him it was his responsibility. The next twenty-four hours showed a remarkable improvement. I couldn't believe how well it had healed; at the end of the trip the doctor told him I'd done a good job of it. I was only too pleased he'd got his two trips in and had a good Christmas.

Several years later when I was skipper of the *Olivean* I was called into the office by Frank Robinson. I was asked to go and see him as he was doing a survey of all the ships' radio equipment. In his office was a maker's model of the *Olivean* in a glass case; I'd admired it many times before. After the survey, on my way out of I stopped to look at it. Frank Robinson came and stood alongside me. As we stood looking at it he turned to me and said, 'Do you know I paid £250 for that model – I tested it in the bath at home and it could hardly stay upright?' 'Well I wouldn't worry about that Mr Robinson because the real thing isn't much better,' I replied. I don't think he was too pleased. 'I think it's time you were going, skipper,' he said. I was only telling him the truth after all!

CHAPTER 26

THESSALONIAN, 1962

I sailed on the *Thessy* on six different occasions. After a couple of months with Len we were getting on really well. The third hand with us was a man getting on in years and he'd been with Len twelve years. He told me he was surprised how much the skipper let me do. Before I came aboard he never trusted anyone to measure anything, he always did it himself. I must have gained his trust, which was great.

Most of my life had been spent fishing for cod and other round fish. I was used to catching lots of fish, but here it was different. Most of the fish we caught was plaice, Dover sole, turbot, brill and any other flatfish. If we caught cod it was only because we ran into them and they were a bonus. I loved cod fishing and when we caught a bag I'd get all excited.

Once when we'd caught a big haul of cod, Len was stood next to me looking at the fish when I picked one up and started stroking it.

'Look at that skipper,' I said. 'That's what a woman's body should be like, firm, smooth and vibrant.'

'You're crazy,' he said.

As he turned round to go back on the bridge I said to him, 'I'd rather have a good bag of cod than a woman any day.'

'And I can bloody well believe you,' he said, shaking his head as he walked away.

Prime fishing was boring. We just plodded about the North Sea looking for them. Ten baskets was a good haul. When Len was in his bunk and I was on the bridge I'd listen to the Ross boats catching cod. When Len rolled out to haul he'd say, 'Don't tell me how much cod the Ross boats are catching because I'm not interested.'

Learning about the prime fishing grounds became really useful to me, and later on when I was skipper I started combining them both. We used to fish about 25 miles east-south-east of Flamborough Head, where the merchant ships would gather before setting out on the convoys. Consequently the place was well mined. The third hand used to keep a daily record on all trips and in the first twelve months he'd recorded that we'd picked up eighteen mines. They were usually rusted and broken, but one or two of them were nearly perfect.

What we did with them depended on the fishing. When the fishing was good we'd lash them up on the port side of the deck, to be dumped later. When the fishing was poor and we were anywhere near a wreck then we'd dump the mine on it and tell the rest of the ships where it was. The proper thing was to report them to the coastguard but they'd tell us to go to the nearest bay, drop anchor and wait for the bomb disposal team. It could take two or three days before they arrived, that was three days' fishing time lost, so we never bothered. If we landed a poor trip it was no good telling the gaffer we'd lost three days because of a mine.

The biggest fright I had with a mine was during a gale. When we dropped the fish on deck there were a mine amongst it. Boy what a beauty, it was nearly perfect. Some of the crew ran aft out of the way. I shouted at them, 'It's no good running away; if this f***ing things goes up, it'll get you wherever you are, so get you arses back here and give us a hand to get rid of it.'

We tied a rope round it and heaved it out on the yoyo, a derrick fixed to the mast which goes outboard and was used for heaving things over the side. I'd already had a word with the man working the yoyo wire so he knew exactly what to do. By now the rest of the crew was well out of the way. I was left with the mine swinging about on the end of the wire. I stood well back and waited for the ship to give a good roll to starboard. A huge sea rolled up and the ship heeled over and I screamed out at the top of my voice, 'LET GO!' There was a hesitation; nothing happened for a few seconds. I was fuming as the ship had now started to roll back to port. Then he suddenly threw the wire off but it was too late. The ship had gone well over to port, and the mine came crashing back onto the deck. When it hit the deck it bounced like a rubber ball. I was sure it'd go off and I called him all the stupid bastards I could think of. After another go we managed to get it over the side. When I asked why he'd hesitated he said his foot was caught in the wire so he'd let go when it was clear. He didn't realise the ship had started rolling back to port.

If we'd done the correct thing and reported it to the coastguard we'd have had to steam 50 or 60 miles in a gale of wind with a mine bouncing about on the deck – I don't think so. Instead we posed for a photo around a mine that could at any minute blow the ship sky high – that's how we were. The danger was only brought home to us all when the Hull trawler *St Hubert* picked one up, as reported in the *Hull Daily Mail*:

St Hubert

The bosun of the Hull trawler *St Hubert*, which sank in the gale-lashed arctic waters off the north coast of Norway yesterday after a metal canister blew up, killing three of the crew and fatally injuring the skipper, said the deck was a shambles after the explosion. The bosun, who took command of the *St Hubert* after the blast, was one of the seventeen survivors taken into Kirkenes, Norway, today by another hull trawler, the *ST Chad*.

Four men, three suffering from shock and one with a slight cut on the head, were taken to the local hospital for treatment. The others, wearing clothes given to them by the *ST Chad*'s crew, drank hot coffee before snatching a few hours' sleep.

The bosun said:
'The object which exploded was like cylinder but with one flat side and was partly covered in concrete. There were fittings which seem to indicate that it had been attached to an anchor and there was a detonator in the nose. We were in a hurry to get home with a good catch in the hold, and stowed the object on the deck', he went on. 'We thought it would be better to dump it in the sea when we were clear of the fishing grounds.

Yesterday with a force 8 gale blowing, I went below after the morning watch and had been down about an hour and it was about 1145 AM when I heard the explosion. I ran up on deck and it was a shambles. The whole port side rail was down, the hatches had been blown open, and the whole foredeck was wrecked. The mate was killed, but his assistant only a couple of feet away was all right.'

The bosun continued:
'There was a terrific squall up to force 10 and I decided to try to run before the gale with the trawler *Prince Charles* keeping near. We kept her going for about 6 hours and then about 5.45 I stopped her and gave the order to abandon ship. She practically sank under us, we used inflatable rafts to get away and go aboard the *Prince Charles*. There was absolutely no panic. Captain Ness was on board the *St Hubert* until she began to sink; we half carried him and half dragged him onto one of the rafts.'

On the trawler were two fourteen-year-old schoolboys on a pleasure trip; luckily both survived. One was the younger brother of the mate, who was killed in the explosion, which, said the owners, was caused by a mine, shell or bomb picked up in the trawler's net. It made us treat mines with a lot more respect after this tragedy.

I stopped to sit for my skipper's ticket and on 17 December 1962 I passed with flying colours and went back in the *Thessy*. The ship Len had just come out of was the *Athenian* – he'd been in her for twelve years. She was one of the old coal burners and had no radar. In foggy weather I always look at the radar scanner to check if it's passing. I hated fog and would rather be fishing in a gale of wind any day.

One foggy day I went on the bridge to take my watch. Len was stood at the wheel being his usual silent self. I stood waiting for him to say something but he didn't, so I asked him if the radar had broken down.

'I don't know,' he said.

'Well, can I put it on?'

'If you want, I wouldn't even know how to switch it on,' he said.

I switched it on and got it tuned in; it was working perfectly. Over the rest of the trip I showed him how to use it. I was the bees' knees after that; any time we had fog I'd go on the bridge and set it all up for him. I must say he picked it up quite easily.

A few trips later we were homeward bound when the fog came down thick and heavy and we were pushing to catch the tide. As we approached the Spurn light vessel Len called me to the bridge. 'I want you to watch the radar; if we have to slow down we'll miss the tide.' I told him I'd watch the radar and let him know what ships were about and where they were. If we needed to slow the speed down I'd tell him. We passed about three cables of the Spurn light vessel; we could hear his fog signal but couldn't see it as the fog was so thick. There were about half a dozen ships or so in the river but we didn't have to slow down for any of them.

At the Lower Burcom light we eased our speed down as we approached the dock gates. The fog was so thick at the entrance to the dock we could hear the men ashore talking but couldn't see them until we were in between the piers. The harbour master told us there was a space just round the corner where we could tie up. 'Don't attempt to go across the dock, the fog's too thick. I'll get a tug to take you to your landing berth in the morning,' he warned us.

After the ship was tied up, Len gave a big sigh of relief. 'Bloody hell,' he said. 'That was nerve wracking; it's the first time in my life I've ever come up the river in thick fog without having to ease down. You did a good job there Jim, well done!'

'First time for me to,' I said.

He nearly had a fit. 'Bloody hell, I wouldn't have done it if I'd known that,' he replied.

We did exactly the same thing the following trip but he was OK, having got confidence in the radar. Len and I really got along, which surprised most people as we were as different as chalk and cheese. At the end of each trip when all the gear was stowed up, he'd call me on the bridge and give me a dram of rum and a can of beer. We'd have a yarn about how much fish we had on board and what gear needed ordering.

That was the only time Len ever drank at sea, a dram and a can of beer at the end of each trip. He always came round to our house on New Year's Day to see the New Year in and have a couple of drams with me.

CHAPTER 27

MT *RHODESIAN*, 1963

I stayed in the *Thessy* till August 1963 then joined the *Rhodesian* on 18 September. The skipper, Johnny Dacombe, was one of the top Faroe/ Westward skippers in the firm. This was new ground to me as I'd no experience of these grounds before.

John was a good skipper to work with. A cockney from London, he came to Grimsby in 1949 at the age of twenty-four and joined Thomas Robinson's. When he came to Grimsby he'd never even seen a trawler, but he soon passed his tickets and became one of their top skippers. He was with them till they ceased trading in the 1970s.

At first I didn't know how to take him (sometimes by the throat may have been a good idea) but he was an excellent seaman, very knowledgeable about seamanship and engines. There weren't many engineers who could tell him anything.

One trip we had a cylinder head pack up on us; he told the engineer to blank it off and we'd run back to Grimsby on five cylinders. The chief said it wasn't possible and they argued about it until John said, 'Right, I'll go and have a call with the engineers in Grimsby.' And John was right: the fitters told the engineer to blank it off and to proceed on five cylinders.

Once I got to know John better we got on very well: in fact we became good friends. When I was at sea my wife spent a lot of time with his wife Maureen. She was good company for her. When we first married we'd decided we'd wait until we'd paid off our loans and got things squared up financially before having children so when my wife found out she was expecting our first child we were very excited.

One day we were laid mending the nets, John leaning out the bridge window watching us. He suddenly shouted to me, 'If that net's on the deck any longer it'll be rotten by the time it goes back in!' When I first sailed with him I panicked at such rebukes, thinking we weren't repairing the net fast enough. Not any more.

'If you hadn't smashed it up in the first place there'd be no need for it to be on the deck at all!' I shouted back. I knew how to take him now!

The following August I was called into the gaffer's office and told John was having three trips off and that I'd be going skipper. I was pleased, although a bit apprehensive. When I arrived on the North Wall to sail one of the runners told me he'd a £2 bet on that I'd not forget to blow the ship's whistle when I went astern, as most new skippers failed to do it. As we got across the dock and headed for the lock gates I could see the runner shaking his fist at me. I realised with a grin that I'd joined the other skippers who'd forgotten.

We cleared the river and passed Spurn light vessel. I set course north by west for Rattery Head. I stood on the bridge looking out at the foredeck; the crew was busy getting the decks squared up. Suddenly I had a feeling of overwhelming pride; this was my dream come true, at last I'd achieved my lifelong ambition to be a skipper. It made all the hard work, the storms, the severe icing and being swilled about the deck from a★★★hole to breakfast time worth every minute.

The elation was short-lived when I went into the skipper's berth and found John had taken all his Decca navigating plots home with him. I was devastated. Here I was on my first trip as skipper bound for the Faroe fishing grounds with no information to go on at all. I bet John's ears were burning as I called him all the rotten sods I could think of. What a lousy trick.

The mate had taken over the watch and I sat in the berth wondering what to do. Not a good start to my first trip as a skipper. I went onto the bridge and sat down, looking out of the window, thinking.

'Are you OK skipper, only you look a bit down in the dumps for someone on his first trip skipper?'

I told him what'd happened with the plots.

'That's no bother. Here take hold of the wheel for me and I'll soon solve that problem.'

He disappeared aft and came back after about ten minutes with two pots of tea and a bag full of Decca plotter rolls. The ship had a plotter worked off a Decca Navigator. The Decca Navigator was a system worked by radio signals; the chart was marked out in different coloured lines, each one numbered and coloured red, green and purple. We worked two colours at a time; if the navigator said you were on the line 20 red and 64 purple you just followed the two lines and where they crossed on the chart was your position. The plotter worked on graph paper with a pen that showed your position, and drew a line following the ship's track. We'd mark out our grounds on these papers, putting in the wrecks and bits of hard ground. They were a big help to us but at times not too reliable.

'There you are skipper, all you want to know about the Faroes and westward fishing grounds.'

The mate had spent plenty of time as a skipper, but was going as mate as he'd had a bad run. He was hoping for another ship of his own in the near future. After this the trip was uneventful and I spent a lot of time copying plots, with his permission, of course. I never got caught like that again and I never asked another skipper if I could copy his plots. As soon as they were asleep I copied everything I could get hold of. Once bitten twice shy as the saying goes.

The fishing was good and we finished up getting a reasonable trip thanks to the mate and one or two other skippers who were willing to help.

The following trip the mate left as he'd been offered a skipper's job with another company. He was down to see us off and wished me well for the future, a true gentleman. After all his help on the previous trip I now felt pretty much at ease as skipper and I soon got into the running of the job.

We went back to the Faroe Islands and it was on the third day fishing that a Danish gunboat arrived on the scene. She was always about, harassing the English trawlers. The Faroe Islands belong to Denmark, hence the Danish gunboats tried their best to get us out of Faroe waters. Her officers came across and inspected our net and measured all our fish. Fortunately we were all legal.

I had a good home life too and everything was going well with my wife and the pregnancy. We didn't know what sex the baby was but it had a good kick, so I reckoned they would be a good footballer. However, during my next trip I received a call from the office and was told the baby had died but my wife was OK. I was so devastated I couldn't speak. I just put the phone down and burst into tears.

I sent the watch to get the mate on the bridge. I had a couple of drams while I waited. When he arrived I told him to take over as I was going to my berth and would be back out when I was ready. I got a couple of cans of beer out the bond locker and gave the mate a bottle of rum to dish out amongst the crew. I went to my berth and left him to it. For about six hours I cried more than I slept, I felt truly devastated.

When I rolled out the mate was drunk. I was totally disgusted and ordered him off the bridge. I've never been so close to killing someone in my life. I didn't speak to him or allow him on the bridge for the rest of the trip. Fortunately it was cut short by the office and we went home.

A couple of trips later we arrived in the River Humber and dropped anchor waiting for the gates to open. We got on with doing the usual jobs on deck ready to enter the dock.

The net store was at the fore end of the ship with a trapdoor in the deck to go down to where the anchor chain was stored. When we heaved the anchor up someone had to go down and keep the chain clear. I went down to the net store with one of the crew, Sid. We opened the lid to find the locker was nearly full of water. It was about 3ft from the top with fish room boards floating on top.

Sid looked down the locker. 'I can get down there easy enough,' he said.

Before I could say anything he jumped into the locker and onto the fish room boards. I watched in amazement as he disappeared under the water leaving his cap floating on top. Bloody hell, I thought, if he gets trapped under the boards he'll drown. I was about to shout for help when he shot up out of the water and I managed to grab hold of him and help him out. 'You idiot, what on earth made you do that?' I said. He told me he thought the boards were all stacked up and didn't realise they were just floating on the top of the water.

He stood there in his best going ashore suit, wet through and shaking like a leaf. I got his cap out of the water and stuck it on his head. 'You'd better go and get dried off and find some gear to go ashore in,' I said.

He got a bit of stick from the rest of the crew as he walked aft, dripping water along the deck as he went.

CHAPTER 28

JUDEAN, OCTOBER 1964/65

I sailed on the *Judean* twice, the first time in October 1964 and the second in September 1965. The first trip was a mixture of good and bad luck. A friend of mine, George Loads, was skipper of the *Ogano* and very knowledgeable about the Faroe Bank. He gave me a lot of information about fishing on top of the bank and round the outer edges into the deep water.

We were fishing on Faroe Bank with a group of ships. Fishing had been good to start with but was steadily getting worse and most ships had left the bank. I decided we'd steam across to the north-east corner of the bank to some of the grounds that George Loads had written down for me. We shot our nets and did a three-hour tow. When we hauled we had a fantastic haul of beautiful large haddock of the very best quality.

We stayed there on our own for the next three days and caught 800 kits of best-quality fish. We went home with expectations of making a big trip. When we landed there was very little fish on the market and prices were sky high. That was the good luck, and all thanks to George!

The bad luck was the fish salesman told me we should've broken the earnings record for our type of ship, but unfortunately the fish was in such a poor state some of it was condemned and sent to the fishmeal factory as unfit for human consumption. This was extremely disappointing as we'd had high hopes of a good payday. The mate got the sack, which was a loss in my opinion as he was a first class mate, but his top priority was looking after the fish.

The second time I sailed in the *Judean* was a bit of a disastrous trip. We sailed on 11 September. We had a young crew who got on very well, and most had sailed with me many times. We had a new crew member, Jock, who was very quiet and didn't have much to say, a bit of a loner. We started our trip at Faroe Bank but the fishing wasn't very good. We spent three or four days there, then left to go and fish on the hundred fathom line west of the Shetland Isles.

On the way across we encountered gale-force winds and stormy seas but by the time we'd reached the 100 fathom line the weather had improved. We shot the nets away in the early afternoon and fished to the eastward during the night. The following morning we hauled at 07.30. The weather was fine, the wind had dropped and we had a smooth sea although with quite a high Atlantic swell. The watch called the crew out at 06.45. I got out of my bunk and went on the bridge. I was looking out the window watching the crew pulling the net in when the mate asked if I'd seen Jock. I'd not seen him. The mate sent one of the crew to go and find him.

'He's not in the toilet and I've had a look round but I can't find him,' he said on his return. I asked if he'd been called out. Apparently he was one of the first out and dressed ready to go on deck. The mate and crew conducted a thorough search of the ship but there was no sign of him. I sent the crew up to all the highest points of the ship to have a good look round. There was no sign whatsoever. I immediately ordered the nets to be brought on board.

I sent out a pan message to Wick Radio giving them our position, explaining we'd lost a man overboard and was doing a thorough search of the area. Unfortunately there were no other ships in the area so we were on our own.

I knew he couldn't be very far away as he could only have gone over the side during the last fifteen minutes. We made a tight search in a 3-mile square of our position, taking into account the direction of the wind and tide. Once we'd covered this area we widened our search. We began to search our way back along the track plotter line on the Decca Navigator. We have the track on the plotter when we're fishing so if we got a good haul we knew exactly which way to go back.

We continued searching for twelve hours. But we could find no trace of the missing man, even though the visibility was excellent and we could see for miles. There were plenty of birds in the area so I told the crew to look out for groups hovering about in one spot. I knew from past experience that if there was anything unusual in the water then the birds tended to hover about above it and make a lot of noise. But there was no sign of anything.

We had to give up the search as there was no way he could've survived after twelve hours in the sea. If he'd been about on top of the water with the conditions as they were I'm sure we'd have found him.

I called the crew into the mess deck to discuss events leading up to Jock's disappearance. Two deckhands that were on watch said they'd called out the crew at 06.45 and Jock was one of the first ready to go on deck. Another said Jock had called him out, saying he was the last man out. When they went on deck, the mate left the mess deck and passed Jock who was stood in the alleyway; the mate carried on to the winch. He was the last man to see Jock alive.

The *Judean* had a variable propeller which controlled the ship's speed; the engines were never stopped. The speed was controlled by altering the pitch of the propellers. The trouble was that if the pitch wasn't set properly when we were supposed to be stopped the ship tended to creep slowly ahead. It was fatal to throw any old net or rope over the side because you could bet your life it would be drawn into the ship's propeller and stop the ship.

In my opinion Jock had left the ship's accommodation on the port side. An alleyway ran along the port side of the ship to just aft side of the bridge where a door led out onto the deck. This is where we found Jock's gear which he'd taken off before going onto the deck. I reckon Jock had gone out onto the port side and somehow gone over the side and been drawn into the ship's propeller. He would've been killed almost instantly and his body would probably have sunk below the waves. After the search we paid our nets away and continued fishing in the area for the next twenty-four hours, still keeping a lookout at all times for any sign of the missing man.

We continued fishing for the next three days and then left the fishing grounds to return to Grimsby. It's a terrible experience to lose a man over the side; you go over it in your mind time and time again wondering if you could've done any more. The atmosphere on the ship was quiet and subdued. I had asked the crew if they wanted to carry on fishing or return to Grimsby but they decided it would serve no purpose to go back early so we continued with the trip. I was sat on the bridge thinking about the things we'd done and if we could've done anything differently, when the bridge door opened. I turned around expecting Jock to walk in, but of course he didn't.

The Board of Trade inquiry had been set for the day we landed. I told all the crew to be down at the Board of Trade, to be well dressed, not drunk and on their best behaviour. I told them I didn't want any messing about as this was an inquiry into a man's death. The inquest was to go ahead at 11.00 and the crew turned up and did me proud. They were all on their best behaviour, nobody had had a drink. Then the bloody mate arrived pissed out of his mind. The inquiry went on for about three hours. We were taken into the inquiry room one at a time and the superintendent asked us questions about the incident.

After the summing up I was asked to enter the room. Gathered in the room were representatives of the firm, the fishermen's union and people from the insurance. Unfortunately there were no representatives of the dead man as nobody could find any relatives, which was very sad. The superintendent came to the conclusion the man had committed suicide for reasons unknown to anybody other than himself. He congratulated me on the good job I'd done in conducting the search; he has also congratulated me on having written such a detailed account of events.

When we returned to the office I wanted the company to sack the mate, but they refused, saying it wouldn't look good, so I had to take him back for another trip. I was pleased the superintendent had given me a good report; at least I felt I'd done all the right things and the best I could.

I left the *Judean* and did three trips in the *Olivean*. We had trouble getting to sea because of the crew – on the first two trips we had to return to port through them getting drunk and fighting, the third time we got away we did a full trip. When we returned to port after the third voyage I was summoned to the superintendent's office. I thought it'd be something to do with the inquiry. I was speechless when he gave me a lecture and threatened to take my ticket off me because we'd not carried out boat drill on any of the three trips. I tried to explain what had happened but he wasn't interested. 'You sailed out of the dock on three separate occasions without doing boat drill,' he told me. 'And if it happens again I will suspend your ticket for three months.' How things change.

CHAPTER 29

SYERSTON, 1966/67

I left the *Judean* in September 1965. While I was at home I received a message from Peter Sleight Trawlers to go down to their office as they'd like a word with me. I was asked if I'd like to go skipper on the *Fiskerton*. I told him I'd have to get permission from Frank Robinson first.

I made an appointment to see Frank Robinson the following morning. He had no objections as he'd no skipper's jobs for me at that moment, but it was on condition I'd return to the firm if a skipper's job came up. This was fine by me, so I joined the *Fiskerton*.

I only did one trip as halfway through I developed appendicitis. I was instructed by the doctor to take the ship back to Grimsby and seek medical attention. When I docked I went to see my GP who said I had a grumbling appendix and that it would settle down and gradually go away.

I reported back to Peter Sleight. He said it wasn't good enough as I was in charge of a ship and men's lives. 'Go and see a specialist. If he says you're OK then you can go back to sea.'

My GP refused to send me to a specialist so I made an appointment to see one privately. On the Friday he confirmed I had appendicitis and been misinformed it would simply go away as I'd been told. He said he'd get me into a private hospital that Monday and they'd operate the next day. After the operation I was told if I'd gone back to sea then within a week the appendix would have burst and I'd have been in serious trouble.

It was a while before I was ready to go to sea again as I had complications after the operation. When I finally went back it wasn't in the *Fiskerton* but the *Syerston*. The gaffer Peter Sleight had been in the Lancaster bombers flying from Lincolnshire air bases during the Second World War so he named all his ships after the air bases he'd flown from: *Fiskerton Syerston, Scampton, Waddington* and *Kirminton*.

The *Syerston* and the *Kerminton* he bought from Aberdeen, the *Kirminton* was originally the *Admiral Cunningham* and the *Syerston* was previously the *Admiral Ramsey*. The *Syerston* was a good ship. She was 110ft long and had a 700hp engine and a variable-pitch propeller similar to the *Judaen*. I liked sailing in her and became very fond of her. Although she wasn't a particularly good sea ship, she was good to me and we always seemed to finish up with a good trip even when the odds were stacked against us.

The first trip we sailed the weather was horrendous with westerly gales and storms. When we left Grimsby we dropped anchor in the River Humber and sorted our fishing gear out. We altered our gear for working hard ground. I didn't know a lot about working off Flamborough Head but I had an old chart that had quite a bit of information on it. Once we had the gear ready we steamed out of the Humber.

The weather was atrocious as we made our way to Flamborough Head, about three hours away. We reached the point where I wanted to start fishing. The wind was still blowing very hard but the sea was quite calm as we were only 3 miles off the land. I was in contact with the ships fishing 40 miles off land. They reported the weather was so bad they had stopped fishing. Well this will do me, I thought. Far better than going out there and rolling about the ocean for nothing! When we hauled our net for the first time we had a decent haul which consisted of plaice, skate, cod and a few Dover sole, turbot and brill. There were about fifteen baskets of fish all told, a good haul.

We continued catching anything from fifteen to twenty baskets, which was a steady living. When I sent our six-day report into the office telling them what fish we had on board the gaffer advised me to come in and land as there was no fish on the market due to the bad weather. We'd been seven days at sea and we landed 200 kits of fish. With the weather being bad and so little fish on the market we made quite a good trip, about £1,900.

The following trip I went back to the same area. The fishing was OK for a couple of days then slacked off, so we moved about the area looking for fish.

It was midday and the mate relieved me to go for my dinner. When I got back on the bridge we were towing our nets alongside one of the Bridlington fishing boats. The mate was stood looking out the window at the vessel. I could see the skipper waving to us, and the mate was waving back to him. Thinking he wanted me on the VHF radio I switched it on and was met with a tirade of abuse. He was calling us all the stupid bastards he could lay his tongue to as we'd just towed our nets through his fishing lines. I apologised, explaining I'd no idea they were line fishing. I asked him to give me the coordinates of the line boats in the area, and where they'd set their lines, so I could keep out of the way. I promised the next time we came to fish at Flamborough Head I'd make contact with the boats before we started fishing and keep clear of their lines.

During the trip I became quite friendly with a few of the Bridlington skippers. They were very helpful and quite a few of them were actually trawling. During the course of the trip I was given quite a bit of information about the grounds. We moved out to the eastward to get away from the line boats. When I told them where I was headed they warned me to be very careful. The ground was very hard, so hard they called it the 'saleroom' as so much fishing gear was lost in the area they were buying fish rather than catching it. I moved into the area to have a go purely because it was good cod ground.

The first couple of tows didn't produce much fish due to the damage we'd done to the net. On the following tow we had some good marks on the fish finder. When we hauled we had a bag of prime cod. There must have been fifty 10-stone boxes. This was brilliant and our next two or three tows produced similar results. When I phoned the office to give our six-day report I said we'd 400 10-stone boxes of fish on board. The gaffer was highly delighted and told us to land the following morning. The weather had not abated for the last couple of weeks and the ships outside were still struggling with the weather and poor fishing. On my way in I contacted my friend Jimmy Lawn, skipper of the *Saxon Progress*, telling him what we'd been catching and where we'd been.

The following trip I returned to the area expecting to find the *Saxon Progress* fishing there, but when I arrived there were no ships in sight. I thought perhaps the fish had taken off and they'd all left the area. We shot the trawl away and had a three-hour tow; when we hauled our nets up we'd a good haul of cod. We fished through the night and caught about ninety 10-stone boxes of cod, an excellent night's work.

Next morning I called Jimmy on the *Saxon Progress*. I asked him what had happened. He told me three ships went to where we'd left. They had twenty-four hours' fishing, and each haul their nets were ripped to shreds and one lost all his gear. I told him I'd shot again and had good fishing during the night, but didn't say how much. He didn't want to know and said I was welcome to it. Later that morning I noticed something on the echo meter like the letter I. Deciding it was nothing to worry about we carried on. About a minute later the ship heeled over to starboard. All the wire started screeching off the winch and then the ship carried on as normal, but at a faster speed, which indicated something was seriously wrong with the gear.

I called the crew out to haul the gear up. When we'd got it up it'd been torn in half, all the wires had parted and the net was ripped straight down the middle top to bottom as if somebody had taken a big knife and slashed right through it. That little line on the echo meter must've been a sunken ship's mast or something similar. It certainly did a lot of damage to our gear.

We did another two or three trips inside Flamborough Head before the fish took off. The weather had improved so we moved off to the other ships that were fishing. We changed our gear over to go fishing for plaice and prime. Fishing was a bit slow and we tried various grounds but couldn't improve it. The Ross Group ships were fishing for cod in the deepwater ground we called the Red House. I heard them talking on the VHF radio about not catching very much fish. Twelve hours later they suddenly went silent, which was a good indication that they'd found fish.

We changed the gear back to the rough ground gear and got in amongst them. Our first haul produced about fifteen to twenty baskets of cod, which was a lot better than we'd been catching, so we stayed with them. They didn't come on the VHF and say what they were catching; they were working on their own private channel. I just kept with them as we were doing OK. I found out later they'd been doing a lot better than us. I tried calling them on the VHF but they just ignored me, so I carried on doing my own thing. After ten days we had a reasonably good trip and went in to land. We landed 350 kits and the fish prices were still good. One of the Ross Group's ships that'd been fishing with us landed 500 boxes that same day and made a lot more money.

When I was in the office talking to Peter Sleight, I asked him if he'd put a variable VHF radio on our ship. He asked if I thought it'd pay off and I assured him it would. I also asked him not tell anybody we'd got it. The following day I went down to the ship as the man from Marconi was there fitting our new radio. I showed him where I wanted it situated. When he'd finished I asked him to show me where the Ross Group's private channel could be found. The following trip I switched the VHF onto Ross's private channel and there it stayed for the rest of the trip and for many more trips after that.

The next trip fishing reports were very poor, so we steamed into the Red House where the Ross Group was fishing. They weren't catching much so we shot the gear south side of them. We were towing north and south and they were towing east and west, crossing my head. They weren't reporting very much fish but the piece of ground we were working on was producing ten to fifteen baskets of good-quality cod each haul so we stayed where we were, but I was listening to them all the time.

It was about 11.00 when I heard one of the Ross boats report thirty baskets of good cod. During the next half hour three more ships hauled up their nets, one reported sixty baskets another had fifty baskets and one forty baskets. This is it, I thought. I was just at the north end of my tow, the other ships were to the east of me, and I pulled my wheel over to starboard and made an easterly course right through the ships. When we hauled I couldn't believe it, we had a huge haul of cod. We had to split the fish three times to get it all on board. Each time we split the fish we hauled in about thirty 10-stone kits of fish and we finished up with ninety to 100 kits of fish. There you are Mr Sleight, your VHF has paid for itself on its first day, job done!

I did this many times; even when I couldn't see them I knew where they were and what they were doing. I often heard them calling hell out of me. Some of them resented the fact I only appeared on the scene when there was fish to be caught. One or two of them would come and speak to me but others didn't. That trip we landed 500 kits of fish – the gaffer was delighted.

On the following trip we steamed out to a ground we called the Easternmost Rough. We'd finished our last trip there and had been catching fifteen to twenty baskets of decent cod so we went straight back there and shot our net. We'd been towing our net for three hours when we caught on an obstruction on the seabed. The net wasn't damaged but there was no fish to speak of. I told the mate to get the gear aboard but changed my mind to give it another go and make sure that the fish really had gone. We towed till just after tea (18.30) and hauled our net. I wasn't really expecting much so was more than surprised to find we had sixty 10-stone boxes of cod. We didn't have a fish washer, so all the fish had to be gutted, thrown across the deck then washed by hand. I told the mate I'd tow the gear till all the fish was gutted, washed and down the fish room. This took four and a half hours. When we hauled again we had the same amount of fish, so we did the same again. We towed till all the fish was down the fish room. We continued doing this till midday the following day. We hauled at 12.30; everyone was tired as the crew had been on deck for almost twenty-four hours. When we hauled we'd 100 boxes of fish. We stopped fishing and lay to while we got it down the fish room.

By now we'd been up almost thirty-six hours so I sent the crew to go and have six hours' sleep. The chief engineer looked after the ship for the next six hours. I told him if any trawler came near, he was to tell them we were laid to because we had a job on the main engine and fishing was terrible.

After our six hours' sleep we started fishing again and then I sent the crew back to bed for another four hours' sleep. During the tow I was listening to the Ross boats; they weren't catching much, so I told them we'd been having some decent fishing, without telling them just how much. I listened to them on their private channel. One of the skippers came on and told the rest of the ships not to believe me, saying I was lying. He told them the ground I was working was fine ground and in all his years at sea he'd never seen cod caught there. Only one ship came to me. He steamed alongside and asked if the fishing was still good. I told him it was so he shot his gear away with us. We fished together for the next twenty-four hours before the fish took off and he stayed off the VHF until the fish went before going back on their private channel to tell the other Ross boats he'd caught 100 kits of fish in twenty-four hours. When they asked why he'd not told them he said, 'You were told twenty-four hours ago and you told everyone Jim was lying, well I can assure you he wasn't.'

We went home that trip and landed 650 kits of cod for six days at sea, all because of a sudden change of mind.

During the '60s and early '70s crews for North Sea trawlers were pretty hard to come by. We used to get all sorts of people signed on the ships. It was the runner's job to get the ship to sea and they'd virtually sign anybody on, it didn't matter if they'd been to sea before or not as long as they got the ship away. One trip we had a new cook. He'd never been on a trawler before but turned out to be an excellent chef. He told us he'd worked in a lot of the big hotels and told us all sorts of stories. One racket was selling joints of meat to the hotel staff. He'd put a joint of meat in a bag for one of the staff and tell him that if anyone stopped him, it was for the dog and unfit for human consumption. When he was making soup he was allowed so many bottles of wine. He would order one bottle for the soup and one for himself. It turned out he was an out-and-out alcoholic. He was with us for quite a few trips. When he came aboard the first two days he was absolutely paralytic, but once he'd sobered up the food was first class and he was an excellent shipmate.

Once we were hauling our nets the crew seemed to be in quite a jolly mood; they were laughing and joking and generally fooling about. When the fish was off the deck and the crew had gone aft I noticed one of them looking a bit unsteady on his feet. I called the mate on the bridge and asked him what was going on. He didn't know, but said he'd noticed a couple of the crew looked as if they'd been drinking. Half an hour later one staggered onto the bridge. He was extremely drunk. I asked where the booze was coming from, as nobody had been the worse for drink since we'd sailed. He said the cook had made a home brew. By six o'clock that evening most of the crew was smashed out of their heads. I went aft with the mate to see if we could locate this alcohol, but we couldn't find it. I called the office, told them what was going on, and that I was bringing the ship back to Grimsby.

I'd seen people drink home brew on several occasions. It's OK while they were drunk and happy, but it makes people very aggressive and they just want to fight. I'd seen it all before and there was no way I was having it on my ship. The gaffer at the office agreed it was the best thing to do. By the time we reached Grimsby half the crew had been fighting and weren't even fit to tie the ship up. I just got my gear, went home and left them to it.

The following trip we had another performance with the crew. In the early hours of the morning I was turned in for a nap, the mate was on watch, and during the tow he got one of the deckhands to relieve him while he went to the toilet.

Next morning one of the skippers told me on the VHF that we were dock-ing on Saturday morning. 'Well it's the first I've heard of it,' I said. Apparently during the night the deckhand had been talking on the VHF to one of his friends, telling him they'd be packing their hands in on Friday morning so they could be in dock to watch Grimsby Town football team play on Saturday. Today was Friday, so I waited to see what would develop during the day. At around 11.30 one of the crew came on the bridge and informed me they'd decided to pack their hands in when we hauled. When we hauled our gear next time they refused to shoot the net away again, saying they were fed up and wanted to go home. I told the mate to tell them they had till 18.00 hours to decide whether they were going to carry on fishing or not. At 18.00 hours they still refused to work. We got the gear on board, put the fish away down the fish room and stowed the net away. We were finished on the deck by 21.00. The mate came on the bridge and asked me what we were going to do. I told him to go aft and tell the crew the skipper had also packed his hand in and he wouldn't be doing anything till breakfast.

When I rolled out at 06.00 on Saturday I went aft and had breakfast. Afterwards I called the mate on the bridge and told him we were going to steam to Grimsby, about eight hours away. High tide was around 15.00 so the gates would open at 13.00, which would give the crew time to get home and changed ready for the match. Or so they thought.

We set off for Grimsby at a steady speed. With high tide being at 15.00 we could get into the dock up to 17.00 before the gates closed. I timed it so we docked at 16.30 in the afternoon. As the crew were leaving they met football fans coming away from the match. They were not pleased, but if they thought I was going to take them in dock in time for a football match after they'd f★★ked up my trip, they were dead wrong. They all got the sack for their troubles.

A couple of trips later we were fishing with the *Ross Falcon* about 60 miles off the River Humber. Fishing was good and we were both doing well. During the afternoon we'd just passed the *Ross Falcon* when we heard a huge explosion. We didn't realise what it was until the skipper of the *Falcon* called us up; there'd been a serious explosion in his engine room.

The skipper was on the bridge when it happened. The second engineer was on watch down in the engine room when he noticed the propeller shaft and bearings at the after end of the engine were quite hot. The chief engineer went down to have a look and told the second engineer to go and fetch a grease gun. He and the chief engineer were on their hands and knees behind the after end of the main engine pumping grease into the propeller shaft when the fore end exploded.

The blast was so fierce it travelled upwards out of the engine room and down the alleyways of the ship's accommodation, blowing three doors off their hinges. It even reached the alleyway to the ladder leading up to the bridge and took the bridge door off. The skipper was stood watching the crews working on deck when it blew his hat off and blew open the bridge doors leading out to the boat deck.

The engine stopped and the skipper gave the order to get the gear up. He called the mate and sent him aft to see what the damage was and what injuries the engineers had sustained. Everyone thought they'd be seriously injured or even killed. The mate went down to the accommodation which was full of smoke, but there didn't appear to be any fire. The mate had breathing apparatus on, and headed to the engine room which was full of black smoke, fearing the worst for the two engineers. It didn't seem possible they'd survived the explosion.

As he got to the ladders he looked down for any signs of life, when out of the smoke appeared the chief followed by the second engineer. Both were completely unhurt, but complained of a loud ringing noise in their ears.

We heaved our gear up and lay alongside the *Falcon*. I was told by the office we were to tow her back to Grimsby. We pulled our warp aft and shackled a chain with a heavy bobbin on it. We got the *Ross Falcon* to pull her warp off and pull it over her bow. I slowly steamed ahead of her till we were both in line, then I went slow astern backing up to her bow. When we were near enough we threw her a heaving line, which was made fast to her warp. We pulled her warp across to us and made it fast to the end of our warp. Then we went slowly ahead. I asked the skipper to slack some of his warp out and once we had enough between us we increased our speed and were making about 7 knots.

All was quite calm and we weren't expecting any bad weather. We'd started towing around 15.00 and by 18.00 we had a full south-west gale blowing and our speed was down to about 4mph.

At about 21.30 there was a terrific bang: we'd parted the tow. Now we had a serious problem, especially with the weather being as it was. Fortunately I'd taken into account that this was a possibility. I had my fore warp laid along the deck with another iron bobbin on it ready just in case. Our warp had parted. We heaved the rest of it onto the winch and the skipper of the *Ross Falcon* heaved his warp up until he came to the shackle which held our warp. There was about 75 fathoms (450ft) of our warp on the end of his. I told him to unshackle it and let it go over the side as it was old warp. When he'd done this we repeated the same manoeuvre as the first time we picked him up. We got in line with him and slowly edged our way back to his bow, a bit tricky as the ship was now rolling and tumbling broadside to the weather. It was very difficult trying to keep level with him as we were driving faster than him.

Once in line we passed him our heaving line. We pulled his warp to us, made it fast to our warp and again we started to tow him. The whole operation from warp parting to picking him up and getting him back on tow took about thirty-five minutes. The skipper paid me a tribute when he said it was the fastest he'd ever seen a tow picked up, and he was a man who'd been at sea a lot more years than I had.

The wind was south-west, so it was on our bow and it made heavy weather of towing. At times our speed was down to about 2 knots. The skipper kept egging me on to give her more speed, but I refused as we were getting along fine albeit slowly and I didn't want to chance parting the warps again.

As we were making towards the Spurn light vessel the weather was easing. Just to the north of the Spurn light vessel there is the Binks buoy which we kept on our starboard side. As we passed the buoy I realised we were being driven down on to it a bit too quickly. The *Falcon* was hanging off our starboard quarter and was closer than I thought. I had to take drastic action or she'd hit it. I put the wheel over to port, came round to an easterly course and dragged her away from the buoy; she only just cleared it. The skipper came on the VHF.

'That was a bit close,' he said.

'Yes, but you didn't hit it.'

'No, but I could've stepped off my ship straight on to it.'

After that it was plain sailing. I informed the Humber authorities I was towing the *Ross Falcon* in and they gave me the all clear to proceed. We made our way up the river to the Burcom Anchorage where a dock tug took the *Falcon* and towed her into dock. Our company sent out some riggers to splice our warps. When this was done we went back to sea to finish off our trip.

There were a couple of skippers who used to take their wives to sea on pleasure trips during the summer months. I asked Pat if she fancied a trip with me. 'I wouldn't mind,' she said. Peter Sleight said it was OK by him and the crew had no problem with it. Some men thought it was unlucky to take a woman to sea on a trawler though. When I told her she was excited, though a bit apprehensive.

We sailed the following day. As we left Grimsby dock the Royal dock was hoisting up a storm cone. That didn't look too promising but I didn't mention it to Pat. By the time we'd steamed six hours to the fishing ground the weather had deteriorated to about force 6 or 7 and we were rolling about quite a lot. Pat had been laid down in the bunk as she'd not been feeling too good, but she came on the bridge to watch us put the gear over the side. When it was down and we were towing along we were head into the wind so the rolling and tumbling wasn't as bad.

I asked the cook if he'd fetch a packet of cream crackers and a cup of water for Pat. I told her to try and keep food down and she'd be OK. After three hours towing the gear, we hauled again. Pat rolled out to watch, looking out the bridge at the crew working on deck.

That haul we had a good bag of cod but the weather was still force 6 to 7, not very nice at all, a case of do we carry on fishing, or not. Owing to the amount of fish we'd caught I decided we'd have another go.

Pat couldn't believe we were going to shoot the gear away again. I explained we often fished in this sort of weather, especially if we were catching a bit of fish.

'But what about the crew?' she said.

'Don't worry about the crew, they know what they're doing and if they thought the weather wasn't worth fishing in they'd soon let me know.'

She was horrified when they went into the fish pound and started gutting the fish when they were still alive. After a couple of days she'd got over feeling queasy and was quite proud of the fact she'd not actually been sick. From then on she enjoyed being on the boat, and took an interest in everything that went on. The weather wasn't fit for her to go on deck, so she watched from the bridge.

When the weather finally calmed and we'd stopped rolling about, Pat went about the ship a bit more. She helped the cook in the galley, doing odd jobs for him. The fishing was quite good for the first five days. After that it slacked off and reports were poor everywhere.

The following day was a beautiful summer's day. The sun was shining, it was warm and the sea was like a millpond. I told Pat we'd steam to Flamborough Head and do some fishing there. On arriving we went straight to the 3-mile limit and shot our gear away. I made contact with the line boats so we could keep clear of them and they told me there'd been some good fishing at the saleroom. We towed the gear out from the land to about 4 miles off, and into the bad ground. This produced some good fishing for the next twenty-four hours.

Next morning the skipper of the *Saxon Onward* told me there was traffic for me at Humber radio. It was a message from the office asking how much fish we had. We had 400 kits so he told me to come in and land the following morning. Pat was a bit disappointed as she was now thoroughly enjoying herself.

We landed 450 kits of fish and made a good trip. Having a woman on board wasn't so unlucky after all. But it wasn't really a good move taking Pat to sea. She did enjoy it, but she also knew what we were going through when the weather was bad, which caused her to worry.

We carried on fishing in the North Sea, and did very well, but nothing much happened. The only incident was when we took a huge sea and it caved our bridge windows in. The water went down the engine room knocking out all the electronics and we had to be escorted back to the River Humber by the trawler *Tiberian*.

We had good cod fishing till the end of March then it slacked off. In April and May fishing in the North Sea was always a struggle. The bigger ships left and went to the Faroe Islands and were doing quite well. When we docked I asked the gaffer if we could try the Faroe fishing grounds.

We did four trips to the Faroe Islands. The first was in March and the weather was atrocious. By the time we arrived the weather had decreased so we shot our gear away and for the first twenty-four hours we fished quite well. The quality of the fish was excellent with a good mixture of cod and haddock. We also caught quite a lot of monkfish, which was classed as rubbish at the time. If we were getting towards the end of the trip and running out of ice all the monkfish would be thrown into a pound without any ice whatsoever. In the 1960s nobody wanted them. Today very high prices are paid for monkfish. We caught lemon sole and quite a lot of medium-sized halibut which made good money.

We'd been on Faroe Bank for three days and the weather was deteriorating. It was blowing force 7 to 8, with a large sea running. With nothing between us and the American coast there was a build up of heavy seas. When we hauled there wasn't much fish. I was debating whether to shoot the net away or bring it on board and dodge out the weather. There were a lot of trawlers still fishing though. I decided to give it another go and see what the weather was like on our next haul.

We were paying our gear away at full speed, and we had the wind on our starboard bow. We'd only been going full speed for a few minutes when the weather deteriorated from a force 7 to a raging force 10. The wind was screaming and we were charging full speed into it. Suddenly a huge sea built up on the starboard bow and broke onto the deck. There were two men on the deck when the sea hit. It filled up the foredeck smashing the deck boards; as I looked out of the wheelhouse window all I could see was a deck full of water and boards, but no men. I honestly thought they'd been washed over the side, I was absolutely horrified!

As the water on the foredeck gradually cleared I saw them. They'd been washed along the ship's rail and were jammed into the bag ropes. They were looking up at the bridge and shouting something at me, but I couldn't hear above the roar of the wind. I don't think I'd have liked what they were saying, but I was so bloody pleased they were OK I didn't give a toss. That was one of the most frightening experiences I ever had. I really thought I'd lost two men overboard.

We got the gear on board, left Faroe Bank and steamed to the Faroe Islands where we fished on what we called the Faroe Flat at the south side of the islands. We landed 400 kits of fish, which wasn't bad considering the weather. Even the bigger trawlers were only taking home around 500 kits. The next trip we landed 688 kits of fish, an excellent trip making a grand total of £2,375.

We did one more trip to the Faroes after that. We landed 410 kits which made £1,820 16s 6d. The gaffer decided we'd go back into the North Sea as fishing had now improved there.

We fished the summer months with excellent results; there were no incidents to write about, just a steady old plod working away in the North Sea.

In the office one trip I was told Sir Thomas Robinson's ships were catching a lot of coleys at the Faroe Flat and I was to go. I was very reluctant as at that time of year, especially September, the jellyfish set in. When they arrive the fish disappear. I told him I wasn't really pleased about going but in the end he talked me round and I went. Another big mistake!

We steamed down to the Faroe Flat; the ships were still fishing there but it wasn't very good. It was on the third day that the jellyfish set in. What fish we'd been catching disappeared and all the ships left the Faroe Islands.

When we arrived on the Papa Bank there were a number of Fleetwood trawlers fishing there. One was the *Josena*, who belonged to J. Marr & Son. George Beech was skipper and I'd known him for years as he was my dad's best friend. They were always out together when they were in dock and he'd often come round to the house with my dad. George's son was also skipper in one of the Marr trawlers. My brother Bill was chief engineer with George, and my youngest brother Ian was doing a pleasure trip with them.

When we shot our gear away it was a lovely summer's day, the sea like glass and not a breath of wind. Once we'd got the gear settled on the seabed I called the mate on the bridge to take the watch, telling him to call me out in two hours time. I'd just got myself settled in my bunk when the ship's whistle sounded. I didn't get out and the mate never called me so I assumed that fog had come down but there were no problems.

The mate called me out when it was time to haul. I rolled out and went to the wheelhouse. As I expected it was dense fog and I looked at the radar to see what ships were in the vicinity. There was a ship just on our starboard quarter, about a quarter of a mile away. The mate said he'd been there most of the tow. He was the only ship on the radar so I relieved the mate and he went to get ready to go on deck.

I sat drinking my mug of tea when I realised I'd not heard anyone talking on the VHF radio. I found the volume had been turned down. When I turned it up the first voice I heard was George Beech senior and he was going ballistic, swearing and shouting at his son, saying his nets were foul with his. George junior insisted it wasn't him and he was nowhere near him. George senior said that some idiot had fouled his gear and he had been towed stern first at 4 knots for the last two hours. He told his son what his position was so just out of curiosity I checked our position. Bloody hells bells! It was us towing him. I called the mate on the bridge.

'Did you see anything of the *Josena* while you were on watch?' I said.

'No, the only ship anywhere near us was the one on our starboard quarter,' he responded.

I called him all the stupid bastards under the sun. That ship, I told him was the *Josena*, and we'd been towing him stern first for the last two hours. I was really annoyed. 'F★★k off the bridge and go get ready to haul the gear!' I don't usually talk to my crew like that but I was livid, I could've gladly strangled him. Now I had to go on the VHF radio and talk to George, and he wasn't a happy bunny at all.

I told George not to heave any of his warps up and I'd sort it out and clear his gear. I was heaving the *Josena* towards me but didn't want her getting too close. Fortunately she wasn't too tangled up and was cleared relatively easily. As we were clearing the gear the *Josena* lay quite close to us and my brothers Bill and Ian came and stood on the bridge veranda. I'd not seen either of them for several months. We were having a chat and catching up on news from home and I heard George telling the other skippers it was like a family gathering with half the Greene clan in attendance!

There'd not been much damage to our gear but as there was no fish we changed grounds. I spoke to Pat on the radio telephone and told her about the fishing being poor at the Faroe Islands. 'I could've told you that. Peter Sleight told me fishing at Faroe had taken off the day after you sailed,' she informed me. I was absolutely furious that the gaffer had just let me carry on going there. We had two or three days' fishing at the Papa Bank on the north of Scotland. We did a lot of damage to the net for very little fish. We were running out of time; we needed to find some fish soon.

We went to the east side of the Shetland Islands but fishing was poor there too. I got onto the office and spoke to Peter Sleight. I wasn't very happy and told him what we were doing. He asked what I intended doing next. 'I don't know, perhaps you can tell me, since you've been running the trip up till now,' I said. I don't think he was too happy. He said he'd make a phone call and get back to me. After five minutes he came back and told me the North Shields ships were working in the North Sea and catching fish so we headed to the grounds they were working. We shot our gear away with them and were catching about fifteen baskets of fish, all quality fish, and nice haddocks with a good sprinkling of lemon sole. We stayed there the rest of the trip but we didn't land a great deal of fish. I certainly must have upset Peter Sleight because he sacked me. I was totally gutted as I was very fond of the *Syerston*.

CHAPTER 30

PORT VALE, 1968

My next vessel after coming out of the *Syerston* was the *Saxon Progress*. She belonged to the Alfred Bannister Fishing Company. They were a small family-owned company and the gaffers were very good to work for.

I left after three trips as I was offered a job in the *Port Vale*, a vessel owned by the Consolidated Fishing Company, or Consols as they were known. When I first came to Grimsby in the '50s it was one of the finest firms in the port. Trying to get onto one of their ships wasn't easy back then and of course they had the pick of the best crews. But now anyone could get on as crews were so short they signed anybody. They had a lot of crew trouble, largely caused by drink. I was asked if I'd take the *Port Vale* to the Faroe Islands. I agreed but told the gaffer I wouldn't be carrying any beer or spirits and to let the crew know.

We sailed on 27 January 1968 and I did three trips in total. When we sailed there was a lot of trouble among the crew who were drinking and fighting. After thirty hours the ship settled down and it emerged that three deckhands had never even been to sea before and the rest weren't great. We had a good mate and third hand, four good engineers, and a cook who should have been certified years ago.

We arrived at the Faroe Bank to start fishing but it took a while to get the gear over the side owing to the inexperience of the crew. When we finally got the gear on the seabed and started fishing we sorted out who was going to do what. I can't say I was pleased with the crew but had to make the best of it.

For the first twenty-four hours the crew got used to hauling and shooting and we weren't doing too badly. The weather was OK and fishing was reasonable. But being the beginning of February the weather wasn't going to stay good for very long and true to form the following day the weather had increased to force 7 or 8. When we were hauling and shooting the gear it took some of the crew all their time to stay on their feet, let alone work the gear. As the net was streaming over the side some crew would be standing on it and in danger of being pulled over the side. They had no idea whatsoever.

We struggled for the next couple of days but the fishing was good so it was worth it. The wind continued to blow force 7 to 8 and the fish took off. Most of the ships had left the bank so we steamed to the south side of the island. Here the weather wasn't much better, but the seas weren't quite as bad as they were on the bank. We arrived just before 18.00 and started fishing again. I was on watch from 18.00 till 24.00. About 20.30 I was sat on the bridge feeling a bit fed up and deep in thought when suddenly the bridge door flew open and in charged the cook. He was waving a carving knife above his head, ranting and raving about the crew and what a load of bastards they were. What he was going to do to them with his knife was nobody's business. Then he disappeared as fast as he came in. I rang the mess room for the watch and I asked the boson what on earth was wrong with the cook; I told him what'd happened. He said not to worry about it as he often had turns like that, but was harmless. Harmless or not, if I was back next trip he wouldn't be.

The last trip was a disaster. In the space of three weeks, three Hull trawlers sank and the Grimsby trawler *Notts County*, sister ship to the *Port Vale*, ran ashore at Iceland.

The weather was terrible. We were battling huge seas on the east coast of the Faroe Islands. I was on the bridge when I heard the news that the third ship, the *Ross Cleveland*, had gone down. I can't put into words how awful I felt at that moment. It was a mixture of deep sorrow and I felt sick to the stomach; the tears just rolled down my cheeks. It was later that afternoon that I heard the Grimsby trawler *Notts County* had also run aground.

CHAPTER 31

TRIPLE TRAGEDY, 1968

On 11 January 1968 the *St Romanus*, owned by Hamlin's, sank 110 miles north-north-east of Spurn. To this day no one really knows why she sank, although the belief is that a sudden squall sealed her fate. A life raft was found two days later by a Danish fishing boat but there was no sign of the twenty-man crew.

Just fifteen days later the *Kingston Peridot* sank north-east of Iceland. At the time a blizzard was raging with severe icing. One of the last messages received stated: 'We're going to lay for a couple of hours while the crew clears ice from the deck.' Heavy oil slicks were later found on the north-east coast along with a partly inflated life raft; there was no sign of the crew.

On 4 February a port left reeling and in deep mourning for the two lost vessels was further devastated by news that the *Ross Cleveland* had also gone down. She was in Isafjord off the north-west coast of Iceland and became the victim of some of the cruellest weather ever experienced at Iceland. For hours a violent storm had battered the ship and the crew hacked away at the crippling ice which formed on the hawser and superstructure. The battle was in vain and the forces of nature won. The last words from skipper Philip Gay were, 'I'm going over; give my love and the crew's love to their wives and families.' She capsized and sank. The weather was so severe that although other trawlers were nearby in the fjord they were unable to help.

But there was a miracle about to happen. Some three days after the *Ross Cleveland* disaster, Harry Eddom, the mate, was found cowering behind a building on the shoreline. He was alive but suffering badly from frostbite. He told the remarkable story of how he'd escaped by forcing the door of the wheelhouse as she capsized. He stumbled along the ice-encrusted vessel and went into the water. He lost consciousness but came round to find himself in a life raft with another man and a boy. Sadly his companions were to die but Harry lived and returned to sea aboard the trawlers.

The *Notts County* was being guided up Isafjord by the Hull trawler *Ross Cleveland* when she disappeared off the radar. Deckhand Frank McGuinness recalled:

> Conditions out there were the worst I have ever seen. The mast was just one block of ice, it was terrifying. Both our radars were knocked out. The *Ross Cleveland* was guiding us to shelter, when she went down we were on our own. When the vessel ran aground the Icelanders moved in. The gunboat *Odin* braved blinding blizzards and mountainous seas to stand by us. Just one mistake and she too would have grounded.

Despite the dangers and the intense cold the Icelanders managed to put a lifeboat and dinghies alongside the 441-ton *Notts County* and take the men off the boat and to hospital in Isafjord. On the afternoon of 7 February they arrived back in Glasgow where deckhand Gilbert Cook broke the silence of the rest of the crew to tell the whole dramatic story:

> I have been twenty-three years at sea but never before have I seen anything like this weather. We had been cutting and axing ice formations on the deck and had just stopped for a cup of tea when it happened. The trawler ran aground. We did not know where we were because our radar scanners were out of action because of the ice. We had to rely on signals from the trawler *Kingston Emerald* which was near us and whose radar was working, conditions were so bad that no one even knew in which direction the vessel was sailing.
>
> Every hand was on deck axing the ice; the blizzard was so thick we could not see the coast or the mountains even after we had grounded. The first thing we did after we had grounded was to lower a lifeboat and rubber dinghies.

At the inquiry into the vessel's lost mate, Barry Stokes took up the story:

> I went on to the boat deck and helped the crew get the starboard life raft out. When that was launched I went across to the port side when the ice came down on top of me and injured my ankle. I couldn't stand up so I ordered all the men to get the life rafts ready. I dragged myself along the casing handrails, by this time the skipper came out and told all hands not to leave the ship because he had been in touch with the Icelandic gunboat *Odin* which would be at the scene within the hour.

Mr Cook said that one crewman jumped into the lifeboat. But as he did so it capsized. Desperate efforts were made to rescue him before the raft was washed on board by the wind and the sea. Mr Stokes told the inquiry: 'I heard from some of the men that he refused to come out of the raft, he kept telling them "Leave me alone". He then crawled back into the life raft.'

Conditions in the fjord were the worst that deckhand Frank McGuinness, an experienced seaman, had ever seen. The mast was just one block of solid ice and it was a terrifying sight. To make matters worse, when the vessel hit the rocks the lights and heating went off. The ship was in complete darkness until the emergency light was pressed into service. On the bridge the crew huddled together, shivering with cold and fear, with not even enough fresh water available to make them a warming drink. And for fifteen hours the waiting went on.

Relief came in the form of a tiny raft seen ploughing towards them in seas whipped into creamy foam by the gale. Aboard it were two sailors from the Icelandic gunboat *Odin* standing by nearly a mile away. This heroic action by men from a ship which just four years later was to terrorise British fishermen who dared to venture into Icelandic waters during the Cod Wars brought praise from *Notts County*' cook Harry Sharp: 'Those Icelanders were risking death themselves to get us off. The raft could easily have overturned.' Battling against the elements, the Icelanders managed to ferry the men to safety, their ordeal finally over.

CHAPTER 32

BACK TO ROBINSON'S, 1968

When I left the *Port Vale* I went back to Tosh Robinson's firm where I was relief skipper from June 1968 until September 1970 when I was given a regular skipper's job on the *Olivean*. Not the best of ships, I'd been on her several times and never did like her, but I was quite successful in her. Then I got promoted to the *Philadelphian*.

I joined the *Philadelphian* in December 1971; again I did very well and she was a far better ship and fished a lot better, so the firm was quite pleased with my earnings. Everything went well till September 1972 when Frank Robinson appointed a new outside manager. Lo and behold it was Billy Woods. He'd been the outside manager for Northern Trawlers, now called British United Trawlers, BUT for short. It was he who'd sacked me when I was mate on the *Serron*.

Once he joined Thomas Robinson's he systematically sacked every skipper in the firm. Some of them, like my friend Lenny Coultas, had been with the firm since they'd left school. He replaced us with skippers who were being made redundant from BUT as they were reducing the size of their fleet, so quite a few skippers and mates were out of work.

Although I was doing quite well he still gave me the sack on 1 December 1972. And I must say that in all the years I'd been in Thomas Robinson's waiting for a regular skipper's job it was ironic that Billy Woods should join the firm and sack me again.

CHAPTER 33

ROSS TIGER, 1972

By this time the only firms still running, apart from Thomas Robinson's, were H.L. Taylor's and BUT. I didn't fancy Taylor's so I tried BUT. I was lucky and got a job as mate on the *Ross Tiger*. I joined her in January 1972. The skipper was Jimmy Brown, a good skipper who did well. Most of the crew had been with Jim for a number of years, so everyone knew what they had to do. I didn't have to tell them anything, they just did it.

The third hand's nickname was 'Dagger'. He liked a bet on the horses, as I did. Just before the Grand National we put our bets on; I backed Red Rum and Dagger backed a horse called Crisp. The race wasn't due to run till the back end of the trip. Every time we were working on the deck Dagger would go on about his horse Crisp. We'd be mending away at the net and he'd slide up alongside me and say things like, 'It's a shame about Red Rum' or 'Jim have you heard the latest: Red Rum is out of the race' or he'd tell me that he'd gone lame. He just wouldn't let it go. One day I said to him, 'Dagger you can't beat good red rum on a crisp day.' The rest of the crew joined in and had their say and it led to a good bit of banter amongst the crew. Some joined me and said Red Rum would win, and some were on the side of Dagger and Crisp.

On the day of the race the skipper worked the hauling times so we could all watch the race on TV in the mess deck. The day of the race there was a buzz around the ship and everyone was fired up for it. As it happened we'd a lot of fish the haul before the race so it looked like we'd be gutting rather than watching the race.

Just before it was due to be run the skipper shouted out the window telling us the cook had made a brew of tea. He sent us aft for a drink and to watch the race. Everybody crowded into the mess deck. You could feel the tension as we waited for the off.

We were all cheering and shouting as the race started, and as the horses started dropping out or falling as they do in the Grand National things really started to hot up. By the time the horses were on the last few jumps Crisp was so far in the lead it didn't look as if he could be caught; it seemed he was home and dry. Red Rum was still going but was a long way behind. I didn't have much to say, but Dagger was full of it. 'I told you, Greenie, where's your Red Rum now?'

As Crisp jumped the last fence Red Rum had made his way into second place but he was never going to catch him. Suddenly things started to change. Red Rum found a second wind and steadily began gaining ground. Crisp was beginning to look tired and was rapidly running out of steam. The cheering got even louder in the mess deck, with everyone egging Red Rum on, but it still looked like Crisp would win.

Red Rum was like a dog with a bone and he wouldn't give up. As they approached the winning post Crisp was dead on his feet and almost down to a walk, he'd nothing to left to give. Red Rum virtually shot past him on the winning line.

It was the most exciting horserace I've ever watched, and of course the banter amongst the crew during the trip had made it all the more exciting. I didn't let Dagger forget it but I bought him a pint out of my winnings. It was a good example of the camaraderie that existed amongst trawler crews, something that sadly seems to be missing these days.

In the winter we were fishing off the west coast of the Shetland Islands, catching huge amounts of large coalfish. We filled the fish room up with them. The firm ordered us to land the fish in Bremerhaven, Germany, because prices were good there. When we left the fishing grounds we called in to Scalloway, a small port in the Shetland Islands. We needed a few repairs doing before setting off for Germany.

The cook told the crew we'd three or four joints of meat left, as we'd only had a short trip, instructing them to see if they could sell it ashore and get some beer money. The lads went to the pub, found a buyer who gave them a good price, the deal was done and they had a good night in the pub. In Scalloway there was a good place to buy freshly smoked kippers. One guy went to the factory and bought a few for the crew, carrying them in a plastic carrier bag. At the end of the night the buyer went home with his meat and the crew came back with their kippers.

We left Scalloway and the following morning the crew was having breakfast when one of them remembered the kippers were still in the alleyway. He left the mess deck to put them in the fridge, returning a few minutes later laughing. 'Guess what, I've found a bag with four joints of meat in it,' he told us.

Everybody burst out laughing. The bags must've been mixed up in the pub. The man who'd paid a good price for four joints of meat had gone home with four boxes of kippers.

'I hope it's a long while before we go into Scalloway again, I certainly wouldn't want to meet him again,' someone exclaimed. According to the crew he was about 6ft tall and nearly as broad!

We landed our fish in Bremerhaven. We were only there about twenty-four hours so didn't have much of a chance to look round. We set off back to the Shetland Isles for more coalies. As we were coming to the south side of the islands the chief engineer informed the skipper a small job on the engine needed doing, and we'd have to go into the nearest port. Scalloway was closest but when we arrived none of the crew dare go ashore for fear they'd meet the man they'd sold the meat to. They couldn't wait to get out of there. Fortunately the job didn't take too long and when we left it was a great relief to the crew. We did well in the *Tiger*, but there didn't seem to be much chance of going skipper in this company so when I was offered a mate's job on the *Osako* I took it and left.

CHAPTER 34

OSAKO, 1973

I joined the *Osako* in June 1973 and stayed until December 1979. The skipper was Peter Newby whom I'd known for many years. I was mate with him in the *Samarian* for a couple of trips in Thomas Robinson's firm. Pete was well known for fishing in bad weather, getting his crew the nickname of Peter's Penguins. He had no regard for the weather at all. When he did get the gear on board it was long after the rest of the ships.

My first experience of this was when we were fishing on the south side of the Faroe Islands. I was on the midnight till 06.30 watch. The weather was force 7 to 8 and there were about ten or twelve trawlers fishing together at the time. We hauled our nets at 04.00. The weather was still bad but hadn't deteriorated so we shot away again. An hour later it turned really nasty and the other ships started getting their gear on board. I listened to them on the VHF radio. They were having a lot of trouble due to the weather. One ship had taken a huge sea and smashed all her deck boards. Another had smashed her fish washer onto the deck. I thought it was time the skipper was out. I went into his berth and explained how bad the weather was. He said to me, 'Is she handling the weather OK?' 'Yes,' I replied. 'Then keep her going and call me at 06.30.'

I went back on the bridge and continued with the watch. Bloody hell, I thought, when does he stop for weather! The only thing we had going for us was the *Osako* was an excellent sea ship and really handled the weather quite well.

I called him out at 06.30 with a pot of tea. The weather was now gusting up to force 9 or more. He walked across the wheelhouse, looked out the weather-side window and remarked, 'Bit rough isn't it.' That was the under-statement of the year! He relieved me and I went aft for breakfast. Afterwards I got into my bunk hoping to get some sleep but it was almost impossible, we were being tossed around all over the place.

I was called out at 10.30 when told the skipper wanted to see me. 'How much fish did we catch last haul?' I asked. 'We haven't had a haul since you went below,' the crewman replied.

I went on the bridge to see the skipper. He told me he'd been waiting for a drop in the weather to get the gear on board but that it was just getting worse. I looked out the window at the weather. Bloody hell, the last time I hauled in weather like this was in the *Northern Pride*, and we know what happened then. I wasn't looking forward to it one bit. The skipper said we were going to try to get the gear on board.

'Take it easy, I'll leave it all up to you but if I shout "water" get out of the way, don't hang about!' he said.

I went aft and told the crew what was happening. I told them to keep their eyes open, watch for the seas rolling up and if they heard the skipper shout, to get the hell out of the way.

We went on deck and started heaving the warps onto the winch. The skipper got the ship round broadside to the seas and stopped the engine. In bad weather we nearly always had trouble getting the trawl doors into the gallows. On this occasion we had no problems at all, everything went like clockwork. I'd never seen gear come aboard so smoothly and sweetly in such bad conditions. We didn't have much fish so we left it in the cod ends. The whole operation took forty minutes, which was bloody good by any standards. Even the skipper congratulated us on a job well done.

Another time when the weather was bad other ships had their gear on board and we were last to haul as usual. This time we couldn't get the trawl doors into the ship's side. We tried all the tricks in the book but nothing seemed to work. The skipper was getting annoyed and was shouting out of the bridge window, which was unusual for him.

It was a very dark night, but with a faint moon trying to get out between the clouds. I was at the winch trying to get the doors into the gallows, tempers were beginning to fray and everyone was wet and cold. As I turned to look aft I spotted something in the distance. About 500ft away was a huge bag of fish bobbing up and down on the top of the seas. This was why we couldn't get the trawl doors in. The sheer pull of the wind and waves on this huge bag of fish was pulling everything away from the ship's side. I shouted to the skipper, even he'd not spotted it. It took ten to fifteen minutes to get the doors alongside and unclipped from the rest of the net. We proceeded to haul the net in very slowly to the ship's side.

As we got it in we could see how much fish was in the net. It was a big haul. After a lot of struggling we heaved the first bag on board. When we started lifting the second bag the becket round the cod end snapped and the bag fell back into the sea. We had no means of dividing the rest of the fish in the net, and they were beginning to sink. When a huge haul of fish like this sinks it's like having a bag of rocks in your net. The skipper called me to the bridge. 'Cut the net away Jim, the weather's too bad and we're going to lose the fish anyway,' he told me. I was a bit surprised, but in view of the weather I could see his point, it was getting dangerous out there. However, I persuaded him to let me have one more try, if I could help it all that fish was going to be on our deck, not the seabed.

Three quarters of an hour later the fish was on the deck, 250 10-stone boxes, a good haul though unfortunately of small fish. The crew had a few hours' gutting while we were dodging the ship head to wind. It wasn't very pleasant for them but I don't like to see fish wasted – if I can get them on board I will. It was my job.

The fishing was becoming increasingly difficult. The Faroe Islands were owned by Denmark and they wanted us off their fishing grounds. They didn't do it by extending the limits as Iceland had done, but by increasing the size of the mesh, especially in the cod end. We were continually harassed by the Danish gunboats. They came aboard us two or three times during each trip.

The British trawlers had been fishing these grounds for hundreds of years and got on very well with the Faroese people. They didn't want us out but it was all down to the Danish government. The mesh size on the cod ends became so big that fishing was becoming unprofitable. Later they followed Iceland and Norway and extended their limits to 200 miles. That finished the fishing there for good.

One of the Grimsby skippers working the Norwegian fishing grounds told us the Norwegians were doing the same with the mesh size in their waters. He said they'd been working beckets across the back of their cod ends and the Norwegian authorities had passed them. He showed our skipper how it was rigged. Our skipper had a meeting with the gaffers the following morning. It was decided that both the *Yesso* and the *Osako* would give it a try in the Faroe waters.

We sailed the next day and on the way to the fishing grounds we rigged these beckets around our net. As I was rigging them up I could see it was never going to work. The beckets weren't long enough for a start and too close to each other. I told the skipper but he insisted they were OK. We arrived on the south side of the Faroe Islands and started fishing. The first haul had about twenty baskets of fish. When we heaved them onto the foredeck the net looked like a long sausage and was halfway up the mast.

'There is no way the gunboat will let us get away with that,' I said to the skipper.

'It doesn't matter, the gaffer has given us the OK so we'll see what the gunboat says,' he replied.

'He won't say anything, he'll just arrest us,' I told him.

The weather wasn't good for the first eight days and consequently we saw no sign of the gunboat. The fishing was good and we caught around 450 kits of excellent-quality cod and haddock, with a good sprinkling of lemon sole and halibut and several large halibuts weighing around 20 stone a piece. On the ninth day the weather fined away and the fish took off. The skipper and Peter Brown, skipper of the *Yesso*, decided they'd go to the north-west of the Faroe Islands (Magnes) and try our luck there. We arrived about 05.00, the weather was good and the sea was calm, a glorious summer's day. I was sure it wouldn't be long before the gunboat arrived on the scene. Sure enough, a couple of hours later it turned up and made straight for us as we were the nearest ship to her. The *Yesso* was about 2 miles further away. He came alongside us and on his megaphone ordered us to haul our gear immediately. The skipper refused, telling him we'd be hauling in two hours time. The gunboat said they were sending a boarding party over and would be putting two officers aboard our ship. The two officers spent the next two hours checking the skipper's logbook and the size of the fish down the fish room.

British trawlers didn't always go by the book. If we could fish inside the limit and get away with it we would, but we never saved undersized fish. Every trawler that landed in England, whether in Grimsby, Hull, Fleetwood or any other fishing port, was subject to strict checks by the Ministry of Agriculture and Fisheries. The officers measured the size of fish every trip and if caught with undersized fish the fines were extremely high.

After two hours we hauled our gear. I was standing in the fish pounds as the cod ends came on board. The officer on the deck took one look and spoke a few words on his radio to the gunboat. He then immediately returned to the bridge and told the skipper to get his gear aboard as he was under arrest for net violation. We got the gear on board.

One of the officers was on the foredeck measuring mesh sizes on the cod ends. His English was very good and he asked me questions about the nets. He reached out and grabbed one of the beckets from around the net. He was about to say something but I jumped in before him.

'Do you know what those beckets are for?' I asked.

'No, but I think you're about to tell me,' he replied.

'Those beckets are strengtheners. When we get large hauls of fish they prevent the back of the net bursting open,' I said.

'Do you know what my job was before I joined the gunboat?' he said.

'No,' I replied.

'I was ten years mate on a fishing trawler,' he said.

'I can't tell you anything then, can I?' I said.

'No,' he replied, grinning at me.

We were escorted to Thorshaven, the capital of the Faroe Islands, to stand trial.

I was on the bridge steaming in and the skipper had gone aft for dinner when the *Yesso* called us up. I explained we were on our way into Thorshaven.

'OK I'll see you when you come out,' he responded. He'd seen the gunboat alongside us so knew what was going on.

The gunboat escorted us into Thorshaven. As soon as we were in the harbour and moored alongside he turned straight round and headed back out to sea. He's going after the *Yesso*, I thought, but knowing Peter Brown he'd be well on his way to the Papa Bank.

The Faroese officials told the skipper he'd be in court at 10.00a.m. the following morning. After tea we sat in the mess deck, the skipper came down and we had a game of poker. The crew was taking the mickey out of him saying he'd be going to jail but told him not to worry as they'd fetch him tobacco and cigarettes to the jailhouse. Peter took it all in good fun.

The next morning he was taken to the courthouse. While we were awaiting the outcome, I decided to go for a walk ashore and have a look around the town. As I walked back to the ship it was all downhill and I had a panoramic view of the harbour and all the ships. I looked at the *Osako* and wondered what on earth the Faroe Islanders thought of us. The ship looked an absolute wreck. She'd not been painted up for months and she looked so scruffy. I can honestly say for the first time in my life I was ashamed to be a crew member of the trawler I was on. The skipper was back on board by the time I arrived and he told me to assemble the crew in the mess deck.

'First of all,' he said, 'I haven't been given a prison sentence, so won't be going to jail. But the bad news is the Faroese have confiscated our fish and are taking it ashore.' The silence was deafening, you could've heard a pin drop. Everyone in the mess deck was absolutely stunned. This was the last thing we expected. We'd had an excellent trip of fish and it was worth quite a bit of money to each crew member. It hit us hard.

About an hour later one of the Faroese officials came on board asking for the mate. I asked him what he wanted. 'Get your crew together and standby to land the fish,' he said. 'If you want that fish landing you're going to have to do it yourself because none of this crew is going to touch it,' I said. After a few cross words he went ashore and returned with a gang of students from the local college, whom they paid to land the fish. The skipper told us when the fish was landed we had to leave the Faroe Islands and go fish on the Papa Bank north of Scotland. The crew told the skipper he could take the ship back to Grimsby. The skipper phoned the office and relayed the message from the crew. The office promised if we did a trip on the Papa Bank they'd pay us top price for the fish we'd lost in Faroe on top of the trip to the Papa Bank. The crew agreed and we spent nine days fishing on the Papa Bank. We spent a total of twenty-one days at sea that trip.

While the fish was being landed a large halibut weighing about 30 stone was heaved from the fish room on the Gilson wire. One of the crew had a Polaroid camera so the skipper took a photograph of it. On the back of the photo he wrote, 'The one that got away'. He gave the photograph to the gaffer, who didn't find it very funny and threatened to sack him.

CHAPTER 35

SPYING ON BRITISH TRAWLERS

It was while I was on the *Osako* that another disaster shook the city of Hull. The Hull trawler *Gaul* was fishing 70 miles north of Norway when on 8 February 1974 she disappeared with all thirty-six crewmen on board. There was a lot of speculation as to how and why she sank. I won't go into details as there were many inquiries and it's well documented, but one theory was that she was spying on the Russians for the British Government. The government denied it but a few years down the line they finally admitted some British trawlers did spy for the them.

I can give you one instance when a trawler from Grimsby was requisitioned by the government to spy along the Russian coast. It was the *Northern Duke* and I know this because my dad was skipper at the time, which was about 1957. He was called into the office, introduced to Commander Brooks, and my dad agreed to take him on a trip to the Russian coast. The office told him to forget about fishing and go wherever Commander Brooks wanted him to go. It was supposed to be top secret, but I believe it was all around the pubs in Grimsby before they'd even left dock!

Dad told me Commander Brooks had thousands of pounds' worth of spying equipment with him. He had a pair of binoculars he could hold out of the bridge window and pick up radar stations. Another time he told my father seven Russian destroyers had just left port and were heading their way and they'd be with them in about an hour and a half. He was correct. Within an hour seven Russian destroyers were sighted heading their way. They weren't particularly making for the *Northern Duke*, they were just passing. They passed down the starboard side of the *Northern Duke* but the last destroyer altered course and headed towards them. All hell broke loose. Commander Brooks dragged all his equipment onto the port side bridge veranda. He told my father there was no way he could be caught with this equipment on the ship. The Russian destroyer proceeded towards the *Duke* but for some unexplained reason he turned around and steamed back to join the convoy and they carried on Westward.

What annoyed me about it all was the fact none of the crew was told what was going on. If the Russians had boarded and found all that equipment on board, she'd have been escorted to Russia. Dad and the crew would've been put on trial and most probably sent to prison. My dad and Commander Brooks would certainly have been found guilty, but the rest of the crew would have spent years in jail doing hard labour through no fault of their own.

Early in 1975 my wife became pregnant again. We were both highly delighted and looking forward to the birth. On 1 December I was working down the fish room when one of the crew called out from the deck that the skipper wanted me on the bridge. It was unusual for him to call me out of the fish room when we were busy. He handed me a telegram which read 'Mother and twins doing well'.

Pete was always one for playing practical jokes. 'Now come on Pete you can do better than that,' I said. I screwed up the telegram, threw it out the window and went back down the fish room. I knew it was a practical joke because the baby wasn't due for another six weeks, and twins indeed!

I hadn't been down the fish room for more than ten minutes when I was summoned back to the bridge. I was fuming. I was getting a bit fed up with this as it was interrupting my work. When I arrived on the bridge the skipper was sat in the radio room. As I went in he said, 'Listen to this'. He picked up the radio telephone and said to his wife, 'Audrey, how is Pat doing?'

'She is OK,' came her reply. 'And the twins are lovely. I've been to see them this morning and they're beautiful.'

I stood there not knowing what to say, I just couldn't believe it. All sorts of things were running through my mind. I couldn't believe we had twins. We'd only catered for one baby and here I was father to a beautiful boy and girl.

The skipper gave me a bottle of rum which I shared amongst the crew then I went back on the bridge. We had another large dram of rum and a couple of cans of beer to wet the babies' heads. I was made up. I couldn't wait to get home to see my wife and the twins.

When we docked Pat and I went to the hospital to see the twins. They were still in incubators and were so tiny and delicate it brought tears to my eyes. One weighed 4lb 3oz and the other just 3lb 10oz.

In August 1978 we'd been fishing at the Faroe Islands; the fishing had been very good and we landed 1,200 kits of excellent-quality fish. We made £21,500, record earnings for one of H.L. Taylor's ships. Both the skipper and I received a bottle of champagne to mark the occasion. As I write this I still have that bottle of champagne in the drinks cupboard. I am not sure it'll be fit to drink now though.

The following trip the skipper had a break, so I took her away as skipper. We steamed to the Faroe Bank and fishing was quite good. By the time the fish had taken off we had 450 kits of good-quality fish. Most of the ships had steamed to the Faroe Flat at the southernmost tip of the islands. They were reporting twenty-five to thirty baskets of good fish. We got our gear aboard and steamed to join them. For some reason, it seemed we weren't catching the same amount of fish as the others.

The tides on the south side of the Faroes are pretty strong. One haul we had a small mending job in the fore part of the net. By the time we had it mended the tide had taken us well to the eastward of the ships. I weighed up the tides and decided not to bother going back as we'd be towing head to tide. I shot the gear away and carried on to the east edge of the Flat. When we hauled we had about sixty baskets, mostly coalies with a sprinkling of cod.

About a mile to the north lie another area we often worked and was usually quite good for cod. Going by the amount of cod we'd caught already I thought it'd be a good bet to head there. When we hauled we had another fifty or sixty baskets of fish, mostly coalies. I shot the net away again, and went a bit more to the eastward. Just before we hauled we came across a small pinnacle on the seabed. There seemed to be quite a lot of fish marks on it so I towed through it and we hauled. We had 150 baskets of fish, which was excellent.

I told the other ships we'd been getting forty or fifty baskets of coalies. They didn't come to us as twenty-five to thirty baskets of cod was a better haul than forty or fifty baskets of coalies.

The following tow we kept to the east end of the tow. When we hauled we had four bags of coalies, about 120 kits. Over the next three days we put 1,000 kits of fish down the fish room. I reported this to the office and they told me to come to Grimsby and land the following Wednesday.

We landed and turned out 1,500 10-stone boxes of fish, a record amount caught by one of the firm's ships, and beating our last trip by about 300 kits. The markets were excellent and I was sure we'd break the earnings record again. When I went down to the office they told me we'd made £21,250 which was £250 short of the record. I was gutted. I always say to this day we probably did break the record, but let's face it they can't have the relief skipper breaking the firm's record. Or maybe it was just sour grapes. Who knows?

The following trip I took her away again; we went back to the Faroe Islands and again the fishing was good. We finished up with a total of about 900 kits and were told to land on the Thursday.

We were due to leave the fishing grounds the following morning and the weather had been fine all day. The forecast was north-west force 6 to 7, ideal as our course for Dennis Head was about south-east by south, which meant we'd be running before the wind and would have an easy trip across.

I rolled out about 03.30 that morning. The mate was on watch and I stood on the bridge while we hauled. We had a reasonable haul, but I wasn't happy, I was on edge. There were times when you'd get the feeling that something wasn't as it should be. I was anxious and eager to be underway. But there was nothing unusual going on, the weather was still fine. We were due to have this next haul and go.

'I've a good mind to get the gear aboard and go now,' I said to the mate. He said it'd be worth having another go so I left him to it and rolled in. I awoke around 05.45. The ship was rolling about a bit. I went on the bridge. It was now blowing pretty hard and I didn't like the look of it one little bit. I told the mate to get all hands out and get the gear aboard as quickly as possible. The wind was now to north-west about force 6–7.

The weather was freshening up and by the time we had the gear on deck and lashed up it was really bad. There wasn't much fish so I told the mate to dump it and get the lads off deck. I rang the engines on to full speed and set our course to Dennis Head. By 11.00 the wind had come round north north-east and was blowing storm 10. The seas were horrendous and the spray was lashing across the ship, driven on by the sheer force of the wind.

I stayed on the bridge watching progress as we made our way across. I could see we were driving slightly south of our course line but I didn't want to alter our course any further east as the sea would be more on our port quarter and we'd likely start taking heavy seas down the port side. The *Osako* was a good sea ship and was running well, not taking much water. I decided to keep the same course and head towards Dennis Head, hoping the weather would be better by the time we got across.

When we got about 20 miles north-west of Dennis Head the weather was north north-east storm 10–11. We were pushing our luck and I made the decision to turn the ship round head to wind and ease the speed down. I called up the office, told them the situation and that there was no way we'd get to Grimsby to land on Thursday.

'Well, when are you going to dock?' the gaffer asked.

I told him it was looking like Thursday evening for the Friday market. He made it clear he wasn't happy, and told me I'd made 'A right balls up of it' as he put it.

I was absolutely furious. Here we were dodging in a north north-east hurricane and all he was worried about was when we were landing. I got off the radio as quick as I could before I said something I'd have later regretted.

We spent the next twelve hours trying to get around Dennis Head. One of the Ross boats who'd been steaming along behind and dodging with us had been told by his office we couldn't make the tide, and if he could make the tide he could land on Thursday. The skipper told them he wouldn't even consider trying.

Once we got round Dennis Head and steamed towards Grimsby things got a lot better. The weather had eased down considerably to a force 7 to gale 8 and we docked Thursday evening for the Friday market.

The following morning I went down to the office, was told our fish had turned out well and the trip had been quite successful. When I'd been paid the gaffer wished to see me in his office. When I went in he shook me by the hand and said, 'Nice trip skipper, you got out of a bad situation quite well there young man'. 'If you're talking about us missing the Thursday market, then I can tell you under the same circumstances I'd do it again, after all it's my responsibility to see that the safety of the ship and crew come before anything else.' His reply: 'Rightly so, off you go and have a nice time in dock skipper. See you next trip.'

The following trip Pete was back and I was back mate with him. It was five years since I'd joined the ship with him and enjoyed every minute of it.

A couple of trips later, I was asked by Henry Taylor if I'd like to go skipper on the *Hondo*. She was smaller than the *Osako* but nevertheless was still a good ship.

We sailed at the beginning of November and the day we sailed we had a westerly gale blowing. The fishing gear was rigged up for fine ground fishing, plaice and sole etc. I decided we'd go and fish off Flamborough Head and work on the edge of the rough ground. As far as I knew no trawlers had fished there since my friend Len came out of Thomas Robinson's.

We arrived at the position, the weather still gale force 8, but we were close enough to land to get a lee, and it was good enough for us to work in.

I was reasonably surprised with our first haul as we'd a good mixture of plaice and roker. The first two trips we landed over 300 kits of plaice. On the third trip we were fishing with one of the firm's other ships. The skipper asked me to go on the VHF radio and work the office channel. I switched channels and skipper started telling me about an incident that had taken place while he was in dock. He was calling hell out of the gaffers. I had to agree with some of the things he said, but wasn't very happy with most of it. I didn't think we should be having this conversation on the private channel, so I made an excuse and ended the conversation.

Next day the other ship left for home and when she sailed again there was a new skipper on board. I didn't think much of this as the previous skipper hadn't been doing very well over the last few months, and with all the skippers ashore they'd plenty to choose from.

I did two more trips on the *Hondo*. When we landed on 22 December I went down to the office and was surprised to find out they were changing skippers and I'd be out of work. Not a very good Christmas present.

There was now only one other firm operating, as Sir Thomas Robinson's had ceased trading in 1975. My only option was British United Trawlers.

The situation was hopeless. Waiting for a mate's job was a long wait and I did two trips in four months. The fishing industry was really all but dead. I couldn't decide what to do. I was thinking of looking somewhere else away from Grimsby, but wasn't sure where. Luckily I was offered a job with BUT as mate on the *Ross Jaguar*. It was a great relief as I couldn't survive on the money from the dole.

A couple of months later I met a friend of mine who was skipper with H.L. Taylor's. I mentioned I was surprised to have been sacked out of the *Hondo*. 'I can tell you why you got the sack,' he said. 'One day I walked into the office and the gaffers were all sat round the VHF listening to the conversation between you and the other skipper, who was calling hell out of them.'

It'd crossed my mind but I was never quite sure about it, as the office never gave any reason as to why they'd sacked me. But now I knew why.

CHAPTER 36

ROSS JAGUAR, 1978

I spent the next year and a half as mate on the *Ross Jaguar* with Dennis Speck. It was a good spell and we earned good money. The skipper was experienced on the Papa Bank. The bank spread out over a large area and the shoal water on top was very hard and rocky, with huge pinnacles. Working on top was OK when there was fish about, sometimes you got away with it, but you'd still get your share of net mending. When there was no fish the nets would be ripped every haul so it wasn't worth the effort. The ground to the east of the Papa Bank, although hard, was easier to work. To the west of the bank was deeper water. Working along the west edge of the bank was still hard ground but not quite as bad as at the top. The fish on and around the Papa Bank was usually good-quality cod and haddock, and quite often with big hauls of spur dogs.

On the west side of the bank Dennis had a favourite tow; he called it his ling tow. A gully ran north and south and the best time to work it was between 09.30 till just after dinner so we'd spend the night towing east and west across the north side of the gully ready for shooting at 09.30. We were catching twenty to thirty baskets of dogfish.

When Dennis rolled out at 06.30 he wasn't very happy to see dogfish. He relieved me and I went aft for breakfast, and then rolled in for some sleep. I got up around 10.00. I went to the galley, got two pots of tea, and took them to the bridge. It was a beautiful day, the sun was shining and the sea was as flat as a millpond. I asked Dennis if it was OK for me to stay on the bridge to see how he worked the tow. He didn't mind so I watched as we went about the tow. On the echo meter there were fish marks very tight on the seabed. Dennis told me these marks were dogfish, and we'd tow for another hour looking for ling and hope that we didn't get too many dogfish. Dogfish swim in huge shoals, and if you got in amongst them you could easily tear away the nets with the sheer weight of fish.

As we carried on towing south in the gully we came across some more marks. Dennis pointed them out and told me they were ling marks. He decided to tow another hour and then haul the gear. I went aft to get my dinner and we hauled at 12.30.

At 12.45 I went on the after deck to watch the net coming to the surface. There were one or two dogfish stuck in the meshes of the net, always a sign you've a good few in the net. The bobbins were heaved up and dropped onto the deck. As the bobbins hit the deck the weight of the fish tried to pull them back over the side. We had to lash them down to keep them on deck, but there was no sign of any ling as they usually float to the top There were a lot of bubbles coming up. As the crew started pulling in the net it gradually and very slowly came to the surface. The whole net from cod end all the way up to the ship's rail was absolutely packed with fish, all dogfish and ling. The buoyancy in the ling was just sufficient to bring it to the surface. It was a fantastic sight. Goodness knows how much fish was in there. The biggest haul of fish I'd seen up to now was thirteen bags of cod which is about 390 kits of fish when I was on the *Northern Jewel*.

There seemed a lot more fish in this net as it lay on top of the sea. Then the worst happened. As the air was squeezed out of the ling the net slowly started to sink, cod end first, and gradually the rest of the net followed. It sank till all the weight of the fish came onto the net and bobbins and began to slowly tear away. There wasn't a thing we could do about it. The sheer weight ripped away the net and it sank to the bottom of the sea, heartbreaking to watch and a bitter disappointment to everyone.

For some reason, when I was with Dennis, the only thing we caught was dogfish. You could bet your bottom dollar when I was on the bridge for my watch we'd be guaranteed a bag of dogs. Dennis would roll out and say, 'Can't you ever catch anything else but dogs?' Even the crew started calling me doggy Jim.

One trip when the fishing was slack, Dennis had been a bit fed up so at 22.00 when the decks were cleared I decided to relieve him early. I went to the bridge and told him I'd take over the watch. He didn't have much to say and went and rolled in. He'd only been gone half an hour when we came across a lot of fish marks. I carried on towing in the same direction until I'd passed through them, then I spun the vessel round and headed straight back towards the marks. I did this three times. We hauled at 13.00. When the net came to the surface there were good indications of a big haul and when we got the fish on deck we'd four bags of fish, all large haddock.

I got the net back on the seabed as soon as possible and towed for another two and a half hours. The fish marks were still there and we got another three bags. By the time I called Dennis out at six o'clock I'd filled the decks. I was really pleased with myself and couldn't wait to tell him what we'd caught. During the early hours of the morning the weather had been gradually freshening up, and by now we had a stiff breeze blowing. When Dennis came on the bridge he looked out the window at the fish, shook his head, and said to me, 'Bloody hell Jim what happens if we get another big haul, we'll have nowhere to put them'. I thought he'd have been pleased, but he never said well done or anything. I thought it was worth a dram of rum at least. I walked off the bridge thinking 'miserable sod'!

Over the last few trips Dennis hadn't been doing very well and we made a few poor trips. When we docked we were still surprised to hear the office had given him the sack. Henry Taylor's firm had packed up in June 1980, so there were a lot of good skippers about.

Our next skipper was Peter Brown but I stopped for a couple of trips off. While I was ashore Peter broke the port earnings record for a middle water fishing vessel which the firm was pleased about. I was told the firm wouldn't be giving any of the mates skipper's jobs, as there were too many good skippers ashore. I was disheartened and getting thoroughly fed up with the job. This was the last fleet of side trawlers fishing out of Grimsby dock so there was just nowhere else to go.

A couple of trips later we'd stopped fishing and the weather was very poor. We got the gear on board and lashed everything down and got everything stowed away as we were expecting some very bad weather on the way home. I relieved the watch at 03.00 hours. The weather was a good force 10, gusting force 11. We were steaming east south-east towards Dennis Head. The wind was west north-west with the seas right on our stern. It really was a bad night. There were huge seas rolling past us. We had a couple of small lights on the foredeck – apart from that and the navigation lights we were in total darkness.

I had the front bridge window down and was looking onto the foredeck. The ship was steaming along quite nicely and not rolling about too much on account of the wind being straight up our stern. As I stood there white flakes started blowing over the top of the bridge. I thought it was starting to snow, after all it was January and we'd been having snowstorms on and off, though nothing out of the ordinary.

Suddenly there was a massive crash on the stern and I could feel the stern being lifted out of the water. We'd been hit by a huge sea. What I thought was snow must've been the foam blowing off the top of the sea as it bore down on top of us.

The sea crashed down onto the deck, smashing the lifeboat and ripping up the air vents, which allowed the water to go down into the after accommodation. It came smashing into the bridge doors, which fortunately opened outwards and were quite sturdy. If they hadn't it would've smashed the doors open and got into the bridge, then we'd have been in serious trouble. The three of us on the bridge would probably have been washed out and lost.

The sea overwhelmed the wheelhouse sending huge amounts of water down the funnel. I'd already closed the bridge windows and as I looked out all I could see was a huge wall of water and for a few seconds the bridge was completely underwater. As it passed over the bridge I found myself looking at the back of this mountainous sea as it went along the foredeck. The sudden weight of water started to press to ship's bow down below the sea.

It was horrendous. I really thought the ship was about to run under. I dashed to the telegraph and rang the engineers to stop the engine. As the engine stopped the skipper came running onto the bridge, asking why I'd stopped the ship. 'Look out of the window,' I screamed. He couldn't believe what he was seeing, the whole foredeck was underwater and you could just see the handrails on the bow, the rest of the foredeck was still underwater. 'F★★king hell,' was all he said.

I don't mind admitting I was terrified. It seemed like an age before the ship started shaking and slowly lifted herself out of the water. First the handrails on the bow came in sight and gradually the water ran off the foredeck and back over the side. One thing in our favour was that the ship didn't roll over to one side. She stayed upright through the whole of the episode and we were very fortunate that the sea hit us square on the after end. Had it hit us on either side I'm quite convinced it would've rolled us right over and we'd have sunk.

Once the water had cleared skipper rang the telegraph for the engineer to go at half speed. About ten minutes later the chief engineer came on the bridge. He was absolutely soaking wet and reported the bilges were full of water and the cabins aft were all flooded. He'd started the pumps to clear away the water. Talk about a night to remember: it was certainly one night we'd remember for a long time to come.

Three months later I packed up. I was disenchanted and there was no future in fishing any more. I'd been thinking hard about it, but what could I do? I had no skills. All I'd ever known was fishing and I was very reluctant to leave.

On 12 December 1980, at the age of forty-two and with a heavy heart, I signed off the *Ross Jaguar* and walked out of BUT's office.

I spent the next three years doing a mobile fish round. I enjoyed it. It was good to meet the general public and for those few years it went well until a lot of other people started doing the same thing. I'd arrive at my pitch only to be told someone had been there two days earlier selling cheap frozen fish. My customers kept asking why I couldn't sell my fish at the same price. I could see it wasn't going to last so I sold it while I still had a business to sell.

CHAPTER 37

ROSS JACKAL, 1984

I returned to fishing and joined the *Ross Jackal* on 18 November 1984. She was one of the lucky ones still sailing and I spent nine months on her. There were about ten of these 'Cat Class' boats still running but they were the last ships left in the port. The others had all been scrapped or sent to the oil rigs as standby boats. Our earnings were quite good as the demand for the fish was there owing to the fact there were so few trawlers left in Grimsby.

On our best trip we landed 1,500 kits of good cod and haddock, and made a total of £56,900. This was the best trip I ever made in a side winder. As time went on the firm laid another couple of the ships up until we were the only trawler left.

The trip prior to our final trip, I had a run-in with one of the customs officers. As we sailed into the dock the man on the lock gates let us know the customs officers from Hull were waiting for us. As we were tying up to the fish market the customs men jumped on board. Three went on the bridge, four went aft, and one stayed on the foredeck. I was on the bow tying up and clearing the ropes. As I came down the ladder onto the foredeck the customs officer was on his knees going through one of the crew's bags. I was absolutely furious. 'And what the f★★k do you think you're doing?' I asked him. The officer turned round to face me, and I was surprised to see a female customs officer. She must've been one of the first. I'd never seen a female customs officer before. I demanded to know what she thought she was doing. 'I'm doing my job checking the crew's bags,' she replied. I really went to town on her.

'Don't you know when the ship comes in dock on her next trip all of these men are going to be out of work? And the only thing they've got to look forward to is being on the dole for months, possibly years. And you've got the cheek to go through all their bags looking for a few fags. Why don't you go on the Royal dock and catch somebody worthwhile who might be smuggling drugs or some other vile stuff ashore. The only thing you're going to get out of this crew is perhaps the odd hundred fags or a couple of packets of tobacco, while in the Royal dock there might be somebody there unloading drugs or hundreds of thousands of cigarettes.'

'And who are you are?' she asked me.

'I'm the mate,' I told her.

'And where are your bags?'

'They're in my berth aft.'

'And where are the keys?'

I told her they were aft and the watchman had them.

'And can I go in your berth?' she asked.

'You can but make sure you leave it as you found it, and give the watchman the keys before you go. By the way you won't find anything because I don't drink and I don't smoke. If I did I wouldn't be stood here giving you all at this grief.'

'Don't you worry I'll do that, and if there's anything in your berth I'll find it, that's one thing you can be sure of.'

The following morning when I went down to see the fish being unloaded I asked the watchman if she'd been and searched my berth.

'No she just came straight aft and gave me the keys, she never bothered with your berth at all,' he said.

'Good job because if she'd looked under the drawers she'd have found four bottles of spirits and 500 cigarettes,' I said.

'Then you're a lucky sod,' the watchman said to me.

During the trip I slipped on the deck and hurt my back. It'd been playing up the entire trip but the morning we landed it was even worse. I went to the office and told the ship's runner I was going to see the doctor. The ship's runner came with me. After the doctor examined me he told me I shouldn't go back on the next trip.

'Why don't you go back and take it easy during the trip. After all they can't sack you, as it's the last trip anyway,' the runner suggested.

I told him I couldn't as it wasn't fair on the rest of the crew but he was firm that if I had another fall or strained my back any further it would probably do me more harm than good.

When the ship docked I went down to watch her come in. It was so sad to see the crew coming off the ship with all their gear. The only other people down to meet her were the watchman and the ship's runner. There was nobody from the office. You'd have thought at least one of the gaffers would've been down to see the men come in. But no, that's all the thanks they got for many years at sea, risking their lives to make money for the company and now facing a bleak future in the dole queue. They'd no other skills and there was no redundancy money as we were classed as casual labour.

There was talk of the government retraining the men and helping them find other jobs, but I never heard of one person being retrained for anything. This was it, a sad end to a once proud and great British fishing industry.

On 11 July 1985, the day after the *Ross Jackal* landed, Dolly Hardy, who'd been fighting for fishermen's compensation for many years, called a meeting at Darley's Hotel on Grimsby Road. There was a good turnout with over 150 fishermen in attendance. It was agreed to keep fighting for compensation. A vote was taken and the British Fishermen's Association was formed. Dolly Hardy and Austin Mitchell, MP for Grimsby, have been doing an excellent job. Unfortunately Dolly passed away recently.

As I write this the date is 14 July 2011 and the fight is still going on; it's a national disgrace the way the government treated the fishermen.

When I first came to Grimsby back in 1954 there were 311 fishing vessels registered, mostly deepwater, middle and near water vessels, but it was an aging fleet. Many of them were built in the early 1900s and all were fired by steam and coal. They were in a terrible state and how they passed their annual surveys is beyond me, but they did. There were plenty of ships and plenty of men to crew them.

After the war when the ships were returned to Grimsby the larger ships that fished Iceland, Bear Island and Norwegian and Greenland waters found huge shoals of cod and haddock. The fish was so plentiful that the crews chopped their heads off to get more into the fish room. Towards the end of the 1940s the owners started to scrap the old coal-burning ships to build larger modern oil-fired steam trawlers. There were more ships being scrapped than being built.

By 1963 the fleet was down to 267 ships, ninety-three of them deep-water ships. With the loss of so many ships the crews struggled to find work and they drifted away from the fishing industry and went to work in the many fish and food factories in Grimsby. With the drift of men leaving things turned around and by the mid-'60s crews became hard to find. By the early '70s even deep-water trawlers were having difficulties and the firms signed on anyone they could.

The biggest loss to the industry was the loss of grounds due to the Cod Wars. The fisheries limit around Iceland was 3 miles and followed the contours of the land; this was good as we got shelter off the land and could continue fishing in bad weather.

In 1952 the Icelandic government extended its limit line out to 4 miles. On 1 September 1958 the limit line was again extended out to 12 miles, a devastating blow as the new limit line didn't follow the contours of the land but went from point to point and cut off hundreds of miles of good fishing grounds.

On 1 September 1972 the second Cod War started. The limit was again extended, this time out to 50 miles. This was serious as it pushed us right out away from the land and shelter. If the weather was bad it was at least 50 miles to get any shelter, and another 50 miles to get back out to the fishing grounds, a 100-mile roundtrip forcing skippers to ride out the storms.

The British Government said they wouldn't accept any extension of the limit but in 1973 they did. On 15 October 1975 Iceland announced it wanted to extend the limit further to 200 miles and the third Cod War began.

On 11 February 1976 the Secretary General of NATO told newsmen in London the USA was extremely interested in the Icelandic situation, especially in view of the importance of the Keflavík Air Base. Mid-1976 and Britain accepted the 200-mile limit. A representative of the British trawling industry declared it a disaster. There was a huge question mark over the survival of Grimsby, Hull and Fleetwood's fishing industries. The representative was right and between December 1976 and December 1978 British trawlers were excluded from the White Sea, Barents Sea, Faroe Islands, the Norwegian coast, Greenland and Newfoundland as other states extended their limits to 200 miles.

All that was left fishing were a few wooden vessels. The BUT middle water cat boats were the last trawlers to fish from Grimsby and were gradually laid up until the *Ross Jackal* landed for a final time in July 1985, marking the end. The industry I loved and which had been my life was gone for good.

In the years that followed I spent three of them running my own fish round (as I mentioned earlier in the book) which wasn't very successful due to fierce competition.

I trained to work on the standby boats which involved a firefighting course and enjoyed the job immensely until it was cut short due to my suffering a heart attack. I had to have a triple bypass operation and though was passed fit after my recovery, I was unable to return to the standby boats due to the strict rules in the industry.

Though I was nearing retirement age anyway I was forced to retire early from work at sea. However, the old *Ross Tiger* was part of the Fishing Heritage Centre in Grimsby and a friend of mine was doing tours on her. It turned out they were looking for another guide and I got the job. It was great, I loved taking the groups round and regaling them with stories of my time on her.

If I could have a pound for every time someone said 'You should write a book' I'd be a rich man by now. And that's really where the idea for this book originated. I enrolled on a computer course as I had little knowledge of computing and bought myself a second-hand PC. With these new skills I started researching and writing the book you have in your hand today.

It's taken a lot of work and dedication but I wanted to share with others my story of a life at sea, a life in an industry sadly forgotten by many, especially our younger generation.

Of course, fishing will always be in my blood and I get out pleasure fishing as often as I can, weather permitting though these days!

I hope you've enjoyed reading my memoirs as much as I've enjoyed writing them.

Jim Greene

Visit our website and discover thousands of other History Press books.

www.thehistorypress.co.uk

The
History
Press